John Tyndall

Light and Electricity

Notes of two Courses of Lectures before the Royal Institution of Great Britain

John Tyndall

Light and Electricity

Notes of two Courses of Lectures before the Royal Institution of Great Britain

ISBN/EAN: 9783337249168

Printed in Europe, USA, Canada, Australia, Japan

Cover: Foto ©berggeist007 / pixelio.de

More available books at **www.hansebooks.com**

LIGHT AND ELECTRICITY:

NOTES

OF

TWO COURSES OF LECTURES BEFORE THE ROYAL INSTITUTION OF GREAT BRITAIN.

BY

JOHN TYNDALL, LL. D., F. R. S.,

AUTHOR OF "HEAT AS A MODE OF MOTION," "LECTURES ON SOUND," "FRAGMENTS OF SCIENCE FOR UNSCIENTIFIC PEOPLE," "HOURS OF EXERCISE IN THE ALPS," ETC., ETC.; PROFESSOR OF NATURAL PHILOSOPHY IN THE ROYAL INSTITUTION OF GREAT BRITAIN.

NEW YORK:
D. APPLETON AND COMPANY,
1, 3, AND 5 BOND STREET.
1893.

PREFACE TO THE AMERICAN EDITION.

For the benefit of those who attended his lectures on Light and Electricity at the Royal Institution, Prof. Tyndall prepared with much care a series of Notes, summing up briefly and clearly the leading facts and principles of these sciences. The Notes proved so serviceable to those for whom they were designed, that they were widely sought by students and teachers, and Prof. Tyndall accordingly had them reprinted in two small books. Under the conviction that they will be equally appreciated by instructors and learners in this country, they are here combined and republished in a single volume.

No intelligent teacher or earnest student needs to be reminded of the importance of repetition and recapitulation to give permanence to mental impressions. But it is neither possible nor desirable to retain in the memory the copious details which may be necessary to the first comprehension of a subject. Hence, after listening to a course of lectures, or going through an extended work in which facts, experimental proofs, and speculations, have

PREFACE TO THE AMERICAN EDITION.

been elaborately presented, it is invaluable to retraverse the field, concentrating attention upon the prominent and established principles of the subject. This is an indispensable condition of all solid acquisition; and, in thus clearly and sharply stating the fundamental principles of Electrical and Optical Science, Prof. Tyndall has earned the cordial thanks of all interested in education.

NEW YORK, *April,* 1871.

CONTENTS.

LIGHT.

	PAGE
General Considerations. Rectilinear Propagation of Light . .	11
Formation of Images through Small Apertures . . .	12
Shadows	13
Enfeeblement of Light by Distance: Law of Inverse Squares .	15
Photometry, or the Measurement of Light	16
Brightness	17
Light requires Time to pass through Space	19
Aberration of Light	20
Reflection of Light (Catoptrics)—Plane Mirrors . . .	22
Verification of the Law of Reflection	22
Reflection from Curved Surfaces: Concave Mirrors . . .	27
Caustics by Reflection (Catacaustics)	31
Convex Mirrors	32
Refraction of Light (Dioptrics)	33
Opacity of Transparent Mixtures	39
Total Reflection	41
Lenses	44
Converging Lenses	44
Diverging Lenses	44
Vision and the Eye	46
Adjustment of the Eye: Use of Spectacles	48
The Punctum Cœcum	50

CONTENTS.

Persistence of Impressions	51
Bodies seen within the Eye	52
The Stereoscope	54
Nature of Light; Physical Theory of Reflection and Refraction	57
Theory of Emission	57
Theory of Undulation	59
Prisms	64
Prismatic Analysis of Light: Dispersion	65
Invisible Rays: Calorescence and Fluorescence	66
Doctrine of Visual Periods	68
Doctrine of Colors	69
Chromatic Aberration. Achromatism	71
Subjective Colors	72
Spectrum Analysis	74
Further Definition of Radiation and Absorption	75
The pure Spectrum: Fraunhofer's Lines	77
Reciprocity of Radiation and Absorption	78
Solar Chemistry	80
Planetary Chemistry	81
Stellar Chemistry	82
Nebular Chemistry	82
The Red Prominences and Envelope of the Sun	82
The Rainbow	84
Interference of Light	86
Diffraction, or the Inflection of Light	88
Measurement of the Waves of Light	93
Colors of Thin Plates	96
Double Refraction	101
Phenomena presented by Iceland Spar	104
Polarization of Light	106
Polarization of Light by Reflection	108
Polarization of Light by Refraction	110
Polarization of Light by Double Refraction	110

CONTENTS.

	PAGE
Examination of Light transmitted through Iceland Spar	111
Colors of Double-refracting Crystals in Polarized Light	114
Rings surrounding the Axes of Crystals in Polarized Light	119
Elliptic and Circular Polarization	120
Rotatory Polarization	121
CONCLUSION	123

ELECTRICITY.

Voltaic Electricity: the Voltaic Battery	131
Electro-Magnetism: Elementary Phenomena	133
Electro-Magnetic Engines	135
Physical Effects of Magnetization	136
Character of Magnetic Force	138
Magnetism of Helix: Strength of Electro-Magnets	140
Electro-Magnetic Attractions: Law of Squares	140
Inference from Law of Squares; Theoretic Notions	143
Diamagnetism: Magne-Crystallic Action	144
Frictional Electricity: Attraction and Repulsion: Conduction and Insulation	145
Theories of Electricity: Electric Fluids	147
Electric Induction: the Condenser: the Electrophorus	148
The Electric Machine: the Leyden-jar	149
The Electric Current	150
The Electric Discharge: Thunder and Lightning	151
Electric Density: Action of Points	152
Relation of Voltaic to Frictional Electricity	153
Historic Jottings, concerning Conduction and the Leyden-jar	155
Historic Jottings, concerning the Electric Telegraph	156
Phenomena observed in Telegraph-Cables	159
Artificial Cables	163
Sketch of Ohm's Theory and Kohlrausch's Verification	165

CONTENTS.

	PAGE
Electro-chemistry. Chemical Actions in the Voltaic Cell: Origin of the Current	168
Chemical Actions at a Distance: Electrolysis	170
Measures of the Electric Current	174
Electric Polarization: Ritter's Secondary Pile	175
Faraday's Electrolytic Law	177
Nobili's Iris Rings	178
Distribution of Heat in the Circuit	179
Relation of Heat to Current and to Resistance	180
Magneto-Electricity: Induced Currents	181
Relation of Induced Currents to the Lines of Magnetic Force. Rotatory Magnetism	184
The Extra-Current	186
Influence of Time on Intensity of Discharge. The Condenser	187
Electric Discharge through rarefied Gases and Vapors	188
Action of Magnets on the Luminous Discharge	190
Magneto-electric Machines. Saxton's Machine. Siemens's Armature	191
Wilde's Machine	192
Siemens's and Wheatstone's Machine	193
Induced Currents of the Leyden-Battery	194

NOTES

OF A COURSE OF NINE LECTURES ON

LIGHT.

NOTES ON LIGHT.

General Considerations. Rectilinear Propagation of Light.

1. THE ancients supposed light to be produced and vision excited by something emitted from the eye. The moderns hold vision to be excited by something that strikes the eye from without. What that something is we shall consider more closely subsequently.

2. *Luminous* bodies are independent sources of light. They generate it and emit it, and do not receive their light from other bodies. The sun, a star, a candle-flame, are examples.

3. *Illuminated* bodies are such as receive the light by which they are seen from luminous bodies. A house, a tree, a man, are examples. Such bodies scatter in all directions the light which they receive; this light reaches the eye, and through its action the illuminated bodies are rendered visible.

4. All illuminated bodies scatter or reflect light, and they are distinguished from each other by the excess or *defect* of light which they send to the eye. A white cloud in a dark-blue firmament is distinguished by its excess of light; a dark pine-tree projected against the same cloud is distinguished through its defect of light.

5. Look at any point of a visible object. The light comes from that point in straight lines to the eye. The

lines of light, or *rays* as they are called, that reach the pupil form a *cone*, with the pupil for a base, and with the point for an apex. The point is always seen at the place where the rays which form the surface of this cone intersect each other, or, as we shall learn immediately, where they *seem* to intersect each other.

6. Light, it has just been said, moves in straight lines; you see a luminous object by means of the rays which it sends to the eye, but you cannot see round a corner. A small obstacle that intercepts the view of a visible point is always in the straight line between the eye and the point. In a dark room let a small hole be made in a window-shutter, and let the sun shine through the hole. A narrow luminous beam will mark its course on the dust of the room, and the track of the beam will be perfectly straight.

7. Imagine the aperture to diminish in size until the beam passing through it and marking itself upon the dust of the room shall dwindle to a mere line in thickness. In this condition the beam is what we call a *ray* of light.

Formation of Images through Small Apertures.

8. Instead of permitting the *direct sunlight* to enter the room by the small aperture, let the light from some body illuminated by the sun—a tree, a house, a man, for example—be permitted to enter. Let this light be received upon a white screen placed in the dark room. Every visible point of the object sends a straight ray of light through the aperture. The ray carries with it the color of the point from which it issues, and imprints the color upon the screen. The sum total of the rays falling thus upon the screen produces an *inverted* image of that object. The image is inverted because the rays *cross* each other at the aperture.

9. *Experimental Illustration.*—Place a lighted candle in a small camera with a small orifice in one of its sides, or a large one covered by tin-foil. Prick the tin-foil with a needle; the inverted image of the flame will immediately appear upon a screen placed to receive it. By approaching the camera to the screen, or the screen to the camera, the size of the image is diminished; by augmenting the distance between them, the size of the image is increased.

10. The boundary of the image is formed by drawing from every point of the outline of the object straight lines through the aperture, and producing these lines until they cut the screen. This could not be the case if the straight lines and the light rays were not coincident.

11. Some bodies have the power of permitting light to pass freely through them; they are *transparent* bodies. Others have the power of rapidly quenching the light that enters them; they are *opaque* bodies. There is no such thing as perfect transparency or perfect opacity. The purest glass and crystal quench some rays; the most opaque metal, if thin enough, permits some rays to pass through it. The redness of the London sun in smoky weather is due to the partial transparency of soot for the red light. Pure water at great depths is blue; it quenches more or less the red rays. Ice when seen in large masses in the glaciers of the Alps is blue also.

Shadows.

12. As a consequence of the rectilinear motion of light, opaque bodies cast shadows. If the source of light be a *point*, the shadow is sharply defined; if the source be a luminous *surface*, the perfect shadow is fringed by an imperfect shadow called a *penumbra*.

13. When light emanates from a point, the shadow of

a sphere placed in the light is a *divergent* cone sharply defined.

14. When light emanates from a luminous globe, the perfect shadow of a sphere equal to the globe in size will be a *cylinder*; it will be bordered by a penumbra.

15. If the luminous sphere be the larger of the two, the perfect shadow will be a *convergent cone*; it will be surrounded by a penumbra. This is the character of the shadows cast by the earth and moon in space; for the sun is a sphere larger than either the earth or the moon.

16. To an eye placed in the true conical shadow of the moon, the sun is totally eclipsed; to an eye in the penumbra, the sun appears horned; while to an eye placed beyond the apex of the conical shadow and within the space enclosed by the surface of the cone produced, the eclipse is *annular*. All these eclipses are actually seen from time to time from the earth's surface.

17. The influence of magnitude may be experimentally illustrated by means of a bat's-wing or fish-tail flame; or by a flat oil or paraffine flame. Holding an opaque rod between the flame and a white screen, the shadow is sharp when the *edge* of the flame is turned toward the rod. When the broad surface of the flame is pointed to the rod, the real shadow is fringed by a penumbra.

18. As the distance from the screen increases, the penumbra encroaches more and more upon the perfect shadow, and finally obliterates it.

19. It is the angular magnitude of the sun that destroys the sharpness of solar shadows. In sunlight, for example, the shadow of a hair is sensibly washed away at a few inches distance from the surface on which it falls. The electric light, on the contrary, emanating as it does from small carbon points, casts a defined shadow of a hair upon a screen many feet distant.

Enfeeblement of Light by Distance; Law of Inverse Squares.

20. Light diminishes in intensity as we recede from the source of light. If the luminous source be a *point*, the intensity diminishes *as the square of the distance increases*. Calling the quantity of light falling upon a given surface at the distance of a foot or a yard—1, the quantity falling on it at a distance of 2 feet or 2 yards is $\frac{1}{4}$, at a distance of 3 feet or 3 yards it is $\frac{1}{9}$, at a distance of 10 feet or 10 yards it would be $\frac{1}{100}$, and so on. This is the meaning of the law of inverse squares as applied to light.

21. *Experimental Illustrations.*—Place your source of light, which may be a candle-flame—though the law is in strictness true only for *points*—at a distance say of 9 feet from a white screen. Hold a square of pasteboard, or some other suitable material, at a distance of $2\frac{1}{4}$ feet from the flame, or $\frac{1}{4}$th of the distance of the screen. The square casts a shadow upon the screen.

22. Assure yourself that the area of this shadow is sixteen times that of the square which casts it; a student of Euclid will see in a moment that this must be the case, and those who are not geometers can readily satisfy themselves by actual measurement. Dividing, for example, each side of a square sheet of paper into four equal parts, and folding the sheet at the opposite points of division, a small square is obtained $\frac{1}{16}$th of the area of the large one. Let this small square, or one equal to it, be your shadow-casting body. Held at $2\frac{1}{4}$ feet from the flame, its shadow upon the screen 9 feet distant will be exactly covered by the entire sheet of paper. When, therefore, the small square is removed, the light that fell upon it is diffused over sixteen times the area on the screen; it is therefore diluted to $\frac{1}{16}$th of its former intensity. That is to say, by

augmenting the distance fourfold we diminish the light sixteenfold.

23. Make the same experiment by placing a square at a distance of 3 feet from the source of light and 6 from the screen. The shadow now cast by the square will have nine times the area of the square itself; hence the light falling on the square is diffused over nine times the surface upon the screen. It is, therefore, reduced to $\frac{1}{9}$th of its intensity. That is to say, by trebling the distance from the source of light we diminish the light ninefold.

24. Make the same experiment at a distance of $4\frac{1}{2}$ feet from the source. The shadow here will be four times the area of the shadow-casting square, and the light diffused over the greater square will be reduced to $\frac{1}{4}$th of its former intensity. Thus, by doubling the distance from the source of light we reduce the intensity of the light fourfold.

25. Instead of beginning with a distance of $2\frac{1}{4}$ feet from the source, we might have begun with a distance of 1 foot. The area of the shadow in this case would be eighty-one times that of the square which casts it; proving that at 9 feet distance the intensity of the light is $\frac{1}{81}$ of what it is at 1 foot distance.

26. Thus when the distances are

1, 2, 3, 4, 5, 6, 7, 8, 9, etc.,

the relative intensities are

1, $\frac{1}{4}$, $\frac{1}{9}$, $\frac{1}{16}$, $\frac{1}{25}$, $\frac{1}{36}$, $\frac{1}{49}$, $\frac{1}{64}$, $\frac{1}{81}$, etc.

This is the numerical expression of the law of inverse squares.

Photometry, or the Measurement of Light.

27. The law just established enables us to compare one light with another, and to express by numbers their relative illuminating powers.

28. The more intense a light, the darker is the shadow which it casts; in other words, the greater is the contrast between the illuminated and unilluminated surface.

29. Place an upright rod in front of a white screen and a candle-flame at some distance behind the rod, the rod casts a shadow upon the screen.

30. Place a second flame by the side of the first, a second shadow is cast, and it is easy to arrange matters so that the shadows shall be close to each other, thus offering themselves for easy comparison to the eye. If when the lights are at the same distance from the screen the two shadows are equally dark, then the two lights have the same illuminating power.

31. But if one of the shadows be darker than the other, it is because its corresponding light is brighter than the other. Remove the brighter light farther from the screen, the shadows gradually approximate in depth, and at length the eye can perceive no difference between them. The shadow corresponding to each light is now illuminated by the other light, and if the shadows are equal it is because the quantities of light cast by both upon the screen are equal.

32. Measure the distances of the two lights from the screen, and square these distances. The two squares will express the relative illuminating powers of the two lights. Supposing one distance to be 3 feet and the other 5, the relative illuminating powers are as 9 to 25.

Brightness.

33. But if light diminishes so rapidly with the distance —if, for example, the light of a candle at the distance of a yard is 100 times more intense than at the distance of 10 yards—how is it that on looking at lights in churches or theatres, or in large rooms, or at our street-lamps, a light

10 yards off appears almost, if not quite, as bright as one close at hand?

34. To answer this question I must anticipate matters so far as to say that at the back of the eye is a screen, woven of nerve-filaments, named the retina; and that when we see a light distinctly, its image is formed upon this screen. This point will be fully developed when we come to treat of the eye. Now the sense of external brightness depends upon the brightness of this internal retinal image, and not upon its size. As we retreat from a light, its image upon the retina becomes smaller, and it is easy to prove that the diminution follows the law of inverse squares; that at a double distance the area of the retinal image is reduced to one-fourth, at a treble distance to one-ninth, and so on. The concentration of light accompanying this decrease of magnitude exactly atones for the diminution due to distance; hence, if the air be clear, the light, within wide variations of distance, appears equally bright to the observer.

35. If an eye could be placed behind the retina, the augmentation or diminution of the image, with the decrease or increase of distance, might be actually observed. An exceedingly simple apparatus enables us to illustrate this point. Take a pasteboard or tin tube, three or four inches wide and three or four inches long, and cover one end of it with a sheet of tin-foil, and the other with tracing-paper, or ordinary letter-paper wetted with oil or turpentine. Prick the tin-foil with a needle, and turn the aperture toward a candle-flame. An inverted image of the flame will be seen on the translucent paper screen by the eye behind it. As you approach the flame the image becomes larger, as you recede from the flame the image becomes smaller; but the *brightness* remains throughout the same. It is so with the image upon the retina.

36. If a sunbeam be permitted to enter a room through a small aperture, the spot of light formed on a distant screen will be *round*, whatever be the shape of the aperture; this curious effect is due to the angular magnitude of the sun. Were the sun a *point*, the light spot would be accurately of the same shape as the aperture. Supposing, then, the aperture to be square, every point of light round the sun's periphery sends a small square to the screen. These small squares are ranged round a circle corresponding to the periphery of the sun; through their blending and overlapping they produce a rounded outline. The spots of light which fall through the apertures of a tree's foliage on the ground are rounded for the same reason.

Light requires Time to pass through Space.

37. This was proved in 1675 and 1676 by an eminent Dane, named Olaf Rœmer, who was then engaged with Cassini in Paris in observing the eclipses of Jupiter's moons. The planet, whose distance from the sun is 475,693,000 miles, has four satellites. We are now only concerned with the one nearest to the planet. Rœmer watched this moon, saw it move round in front of the planet, pass to the other side of it, and then plunge into Jupiter's shadow, behaving like a lamp suddenly extinguished: at the other edge of the shadow he saw it reappear like a lamp suddenly lighted. The moon thus acted the part of a signal-light to the astronomer, which enabled him to tell exactly its time of revolution. The period between two successive lightings up of the lunar lamp gave this time. It was found to be 42 hours, 28 minutes, and 35 seconds.

38. This observation was so accurate, that having determined the moment when the moon emerged from the shadow, the moment of its hundredth appearance could

also be determined. In fact, it would be 100 times 42 hours, 28 minutes, 35 seconds, from the first observation.

39. Rœmer's first observation was made when the earth was in the part of its orbit nearest Jupiter. About six months afterward, when the little moon ought to make its appearance for the hundredth time, it was found unpunctual, being fully 15 minutes behind its calculated time. Its appearance, moreover, had been growing gradually later, as the earth retreated toward the part of its orbit most distant from Jupiter.

40. Rœmer reasoned thus: "Had I been able to remain at the other side of the earth's orbit, the moon might have appeared always at the proper instant; an observer placed there would probably have seen the moon 15 minutes ago, the retardation in my case being due to the fact that the light requires 15 minutes to travel from the place where my first observation was made to my present position."

41. This flash of genius was immediately succeeded by another. "If this surmise be correct," Rœmer reasoned, "then as I approach Jupiter along the other side of the earth's orbit, the retardation ought to become gradually less, and when I reach the place of my first observation there ought to be no retardation at all." He found this to be the case, and thus proved not only that light required time to pass through space, but also determined its rate of propagation.

42. The velocity of light as determined by Rœmer is 192,500 miles in a second.

The Aberration of Light.

The astounding velocity assigned to light by the observations of Rœmer received the most striking confirma-

THE ABERRATION OF LIGHT. 21

tion from the English astronomer Bradley in the year 1723. In Kew Gardens to the present hour there is a sundial to mark the spot where Bradley discovered the aberration of light.

43. If we move quickly through a rain-shower which falls vertically downward, the drops will no longer seem to fall vertically, but will appear to meet us. A similar deflection of the stellar rays by the motion of the earth in its orbit is called the *aberration of light*.

44. Knowing the speed at which we move through a vertical rain-shower, and knowing the angle at which the rain-drops appear to descend, we can readily calculate the velocity of the falling drops of rain. So, likewise, knowing the velocity of the earth in its orbit, and the deflection of the rays of light produced by the earth's motion, we can immediately calculate the velocity of light.

45. The velocity of light, as determined by Bradley, is 191,515 miles per second—a most striking agreement with the result of Rœmer.

46. This velocity has also been determined by experiments over terrestrial distances. M. Fizeau found it thus to be 194,677 miles a second, while the later experiments of M. Foucault made it 185,177 miles a second.

47. "A cannon-ball," says Sir John Herschel, "would require seventeen years to reach the sun, yet light travels over the same space in eight minutes. The swiftest bird, at its utmost speed, would require nearly three weeks to make the tour of the earth. Light performs the same distance in much less time than is necessary for a single stroke of its wing; yet its rapidity is but commensurate with the distance it has to travel. It is demonstrable that light cannot reach our system from the nearest of the fixed stars in less than five years, and telescopes disclose to us objects probably many times more remote."

The Reflection of Light (Catoptrics)—Plane Mirrors.

48. When light passes from one optical medium to another, a portion of it is always turned back or reflected.

49. Light is *regularly* reflected by a polished surface; but if the surface be not polished, the light is *irregularly* reflected or scattered.

50. Thus a piece of ordinary drawing-paper will scatter a beam of light that falls upon it so as to illuminate a room. A plane mirror receiving the sunbeam will reflect it in a definite direction, and illuminate intensely a small portion of the room.

51. If the polish of the mirror were perfect it would be invisible, we should simply see in it the images of other objects; if the room were without dust-particles, the beam passing through the air would also be invisible. It is the light scattered by the mirror and by the particles suspended in the air which renders them visible.

52. A ray of light striking as a perpendicular against a reflecting surface is reflected back along the perpendicular; it simply retraces its own course. If it strike the surface obliquely, it is reflected obliquely.

53. Draw a perpendicular to the surface at the point where the ray strikes it; the angle enclosed between the *direct* ray and this perpendicular is called the angle of incidence. The angle enclosed by the *reflected* ray and the perpendicular is called the angle of reflection.

54. It is a fundamental law of optics that *the angle of incidence is equal to the angle of reflection.*

Verification of the Law of Reflection.

55. Fill a basin with water to the brim, the water being blackened by a little ink. Let a small plummet—a small lead bullet, for example—suspended by a thread,

VERIFICATION OF THE LAW OF REFLECTION.

hang into the water. The water is to be our horizontal mirror, and the plumb-line our perpendicular. Let the plummet hang from the centre of a horizontal scale, with inches marked upon it right and left from the point of suspension, which is to be the zero of the scale. A lighted candle is to be placed on one side of the plumb-line, the observer's eye being at the other.

56. The question to be solved is this: How is the ray which strikes the liquid surface at the foot of the plumb-line reflected? Moving the candle along the scale, so that the tip of its flame shall stand opposite different numbers, it is found that, to see the reflected tip of the flame *in the direction of the foot of the plumb-line*, the line of vision must cut the scale as far on the one side of that line as the candle is on the other. In other words, the ray reflected from the foot of the perpendicular cuts the scale accurately at the candle's distance on the other side of the perpendicular. From this it immediately follows that the angle of incidence is equal to the angle of reflection.

57. With an artificial horizon of this kind, and employing a theodolite to take the necessary angles, the law has been established with the most rigid accuracy. The angle of elevation to a star being taken by the instrument, the telescope is then pointed downward to the image of the star reflected from the artificial horizon. It is always found that the direct and reflected rays enclose equal angles with the horizontal axis of the telescope, the reflected ray being as far below the horizontal axis as the direct ray is above it. On account of the star's distance the ray which strikes the reflecting surface is parallel with the ray which reaches the telescope directly, and from this follows, by a brief but rigid demonstration, the law above enunciated.

58. The path described by the direct and reflected rays is the shortest possible.

59. When the reflecting surface is roughened, rays from different points, more or less distant from each other, reach the eye. Thus, a breeze crisping the surface of the Thames or Serpentine sends to the eye, instead of single images of the lamps upon their margin, pillars of light. Blowing upon our basin of water, we also convert the reflected light of our candle into a luminous column.

60. Light is reflected with different energy by different substances. At a perpendicular incidence, only 18 rays out of every 1,000 are reflected by water, 25 rays per 1,000 by glass, while 666 per 1,000 are reflected by mercury.

61. When the rays strike obliquely, a greater amount of light than that stated in 60, is reflected by water and glass. Thus, at an incidence of 40°, water reflects 22 rays; at 60°, 65 rays; at 80°, 333 rays; and at $89\frac{1}{2}$° (almost grazing the surface), it reflects 721 rays out of every 1,000. This is as much as mercury reflects at the same incidence.

62. The augmentation of the light reflected as the obliquity of incidence is increased may be illustrated by our basin of water. Hold the candle so that its rays enclose a large angle with the liquid surface, and notice the brightness of its image. Lower both the candle and the eye until the direct and reflected rays as nearly as possible graze the liquid surface; the image of the flame is now much brighter than before.

Reflection from Looking-glasses.—Various instructive experiments with a looking-glass may be here performed and understood.

63. Note first when a candle is placed between the glass and the eye, so that a line from the eye through the candle is perpendicular to the glass, that *one* well-defined image of the candle only is seen.

64. Let the eye now be moved so as to receive an oblique reflection; the image is no longer single, a series of images at first partially overlapping each other being seen. By rendering the incidence sufficiently oblique these images, if the glass be sufficiently thick, may be completely separated from each other.

65. The first image of the series arises from the reflection of the light from the anterior surface of the glass.

66. The second image, which is usually much the brightest, arises from reflection at the silvered surface of the glass. At large incidences, as we have just learned, metallic reflection far transcends that from glass.

67. The other images of the series are produced by the reverberation of the light from surface to surface of the glass. At every return from the silvered surface a portion of the light quits the glass and reaches the eye, forming an image; a portion is also sent back to the silvered surface, where it is again reflected. Part of this reflected beam also reaches the eye and yields another image. This process continues: the quantity of light reaching the eye growing gradually less, and, as a consequence, the successive images growing dimmer, until finally they become too dim to be visible.

68. A very instructive experiment illustrative of the augmentation of the reflection from glass, through augmented obliquity, may here be made. Causing the candle and the eye to approach the looking-glass, the first image becomes gradually brighter; and you end by rendering the image reflected from the glass brighter, more luminous, than that reflected from the metal. Irregularities in the reflection from looking-glasses often show themselves; but with a good glass—and there are few glasses so defective as not to possess, at all events, some good portions —the succession of images is that here indicated.

69. *Position and Character of Images in Plane Mirrors.*—The image in a plane mirror appears as far behind the mirror as the object is in front of it. This follows immediately from the law which announces the equality of the angles of incidence and reflection. Draw a line representing the section of a plane mirror; place a point in front of it. Rays issue from that point, are reflected from the mirror, and strike the pupil of the eye. The pupil is the base of a cone of such rays. Produce the rays backward; they will intersect behind the mirror, and the point will be seen *as if* it existed at the place of intersection. The place of intersection is easily proved to be as far behind the mirror as the point is in front of it.

70. Exercises in determining the positions of images in a plane mirror, the positions of the objects being given, are here desirable. The image is always found by simply letting fall a perpendicular from each point of the object, and producing it behind the mirror, so as to make the part behind equal to the part in front. We thus learn that the image is of the same size and shape as the object, agreeing with it in all respects save one—the image is a *lateral inversion* of the object.

71. This inversion enables us, by means of a mirror, to read writing written backward, as if it were written in the usual way. Compositors arrange their type in this backward fashion, the type being reversed by the process of printing. A looking-glass enables us to read the type as the printed page.

72. Lateral inversion comes into play when we look at our own faces in a glass. The right cheek of the object, for example, is the left cheek of the image; the right hand of the object the left hand of the image, etc. The hair parted on the left in the object is seen parted to the right of the image, etc.

73. A plane mirror half the height of an object gives an image which embraces the whole height. This is readily deduced from what has gone before.

74. If a plane mirror be caused to move parallel with itself, the motion of an image in the mirror moves with twice its rapidity.

75. The same is true of a *rotating* mirror: when a plane mirror is caused to rotate, the angle described by the image is twice that described by the mirror.

76. In a mirror inclined at an angle of 45 degrees to the horizon, the image of an erect object appears horizontal, while the image of a horizontal object appears erect.

77. An object placed between two mirrors enclosing an angle yields a number of images depending upon the angle enclosed by the mirrors. The smaller the angle, the greater is the number of images. To find the number of images, divide 360° by the number of degrees in the angle enclosed by the two mirrors, the quotient, if a whole number, will be the number of images, plus one, or it will include the images and the object. The construction of the kaleidoscope depends on this.

78. When the angle becomes 0—in other words, when the mirrors are parallel—the number of images is infinite. Practically, however, we see between parallel mirrors a long succession of images, which become gradually feebler, and finally cease to be sensible to the eye.

Reflection from Curved Surfaces: Concave Mirrors.

79. It has been already stated and illustrated that light moves in straight lines, which receive the name of rays. Such rays may be either divergent, parallel, or convergent.

80. Rays issuing from terrestrial points are necessarily divergent. Rays from the sun or stars are, in consequence

of the immense distances of these objects, sensibly parallel.

81. By suitably reflecting them, we can render the rays from terrestrial sources either parallel or convergent. This is done by means of *concave* mirrors.

82. In its reflection from such mirrors, light obeys the law already enunciated for plane mirrors. The angle of incidence is equal to the angle of reflection.

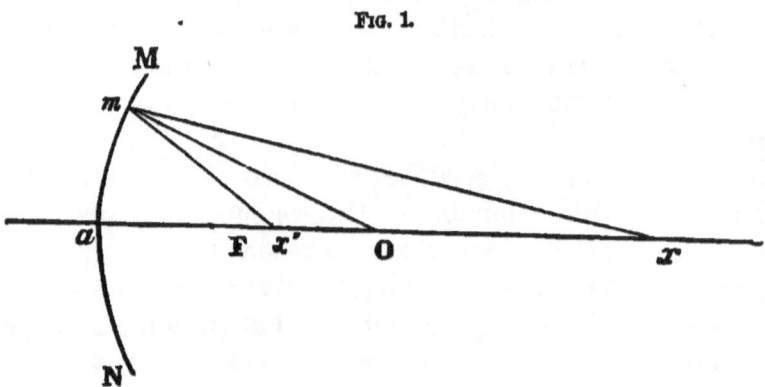

Fig. 1.

83. Let M N be a very small portion of the circumference of a circle with its centre at O. Let the line ax, passing through the centre, cut the arc M N into two equal parts at a. Then imagine the curve M N twirled round ax as a fixed axis; the curve would describe part of a spherical surface. Suppose the surface turned toward x to be silvered over, we should then have a concave spherical reflector; and we have now to understand the action of this reflector upon light.

84. The line ax is the principal axis of the mirror.

85. All rays from a point placed at the centre O strike the surface of the mirror as perpendiculars, and after reflection return to O.

86. A luminous point placed on the axis beyond O, say

at x, throws a divergent cone of rays upon the mirror. These rays are rendered convergent on reflection, and they intersect each other at some point on the axis between the centre O and the mirror. In every case the direct and the reflected rays (xm and mx' for example) enclose equal angles with the radius (Om) drawn to the point of incidence.

87. Supposing x to be exceedingly distant, say as far away as the sun from the small mirror—or, more correctly, supposing it to be *infinitely* distant—then the rays falling upon the mirror will be *parallel*. After reflection such rays intersect each other, *at a point midway between the mirror and its centre*.

88. This point, which is marked F in the figure, is the *principal focus* of the mirror; that is to say, the principal focus is the focus of *parallel rays*.

89. The distance between the surface of the mirror and its principal focus is called the *focal distance*.

90. In optics, the position of an object and of its image are always exchangeable. If a luminous point be placed in the principal focus, the rays from it will, after reflection, be parallel. If the point be placed anywhere between the principal focus and the centre O, the rays after reflection will cut the axis at some point beyond the centre.

91. If the point be placed between the principal focus F and the mirror, the rays after reflection will be *divergent* —they will not intersect at all—there will be no *real* focus.

92. But if these divergent rays be produced backward, they will intersect *behind* the mirror, and form there what is called a *virtual*, or imaginary focus.

Before proceeding further, it is necessary that these simple details should be thoroughly mastered. Given the position of a point in the axis of a concave mirror, no difficulty must be experienced in finding the position of the image of that point, nor in determining whether the focus is *virtual* or *real*.

93. It will thus become evident that, while a point moves from an infinite distance to the centre of a spherical mirror, the image of that point moves only over the distance between the principal focus and the centre. Conversely, it will be seen that during the passage of a luminous point from the centre to the principal focus, the image of the point moves from the centre to an infinite distance.

94. The point and its image occupy what are called *conjugate foci*. If the last note be understood, it will be seen that the conjugate foci move in opposite directions, and that they coincide at the centre of the mirror.

95. If instead of a point an object of sensible dimensions be placed beyond the centre of the mirror, an *inverted* image of the object *diminished* in size will be formed between the centre and the principal focus.

96. If the object be placed between the centre and the principal focus, an inverted and *magnified* image of the object will be formed beyond the centre. The positions of the image and its object are, it will be remembered, convertible.

97. In the two cases mentioned in 95 and 96 the image is formed in the air in *front* of the mirror. It is a *real* image. But if the object be placed between the principal focus and the mirror, an *erect* and magnified image of the object is seen behind the mirror. The image is here *virtual*. The rays enter the eye *as if* they came from an object behind the mirror.

98. It is plain that the images seen in a common looking-glass are all virtual images.

99. It is now to be noted that what has been here

stated regarding the gathering of rays to a *single focus* by a spherical mirror is only true when the mirror forms a small fraction of the spherical surface. Even then it is only practically, not strictly and theoretically, true.

Caustics by Reflection (*Catacaustics*).

100. When a large fraction of the spherical surface is employed as a mirror, the rays are not all collected to a point; their intersections, on the contrary, form a luminous *surface*, which in optics is called a *caustic* (German, Brennfläche).

101. The interior surface of a common drinking-glass is a curved reflector. Let the glass be nearly filled with milk, and a lighted candle placed beside it; a caustic curve will be drawn upon the surface of the milk. A carefully-bent hoop, silvered within, also shows the caustic very beautifully. The focus of a spherical mirror is the *cusp* of its caustic.

102. *Aberration.*—The deviation of any ray from this cusp is called the *aberration* of the ray. The inability of a spherical mirror to collect all the rays falling upon it to a single point is called the *spherical aberration* of the mirror.

103. Real images, as already stated, are formed in the air in front of a concave mirror, and they may be seen in the air by an eye placed among the divergent rays beyond the image. If an opaque screen, say of thick paper, intersect the image, it is projected on the screen and is seen *in all positions* by an eye placed in front of the screen. If the screen be semi-transparent, say of ground glass or tracing-paper, the image is seen by an eye placed either in front of the screen or behind it. The images in phantasmagoria are thus formed.

Concave spherical surfaces are usually employed as

burning-mirrors. By condensing the sunbeams with a mirror 3 feet in diameter and of 2 feet focal distance, very powerful effects may be obtained. At the focus, water is rapidly boiled, and combustible bodies are immediately set on fire. Thick paper bursts into flame with explosive violence, and a plank is pierced as with a hot iron.

Convex Mirrors.

104. In the case of a *convex* spherical mirror the positions of its foci and of its images are found as in the case of a concave mirror. But all the foci and all the images of a convex mirror are virtual.

105. Thus to find the principal focus you draw parallel rays which, on reflection, enclose angles with the radii equal to those enclosed by the direct rays. The reflected rays are here *divergent;* but on being produced backward, they intersect at the principal focus *behind the mirror*.

106. The drawing of *two* lines suffices to fix the position of the image of any point of an object either in concave or convex spherical mirrors. A ray drawn from the point through the centre of the mirror will be reflected through the centre; a ray drawn parallel to the axis of the mirror will, after reflection, pass, or its production will pass, through the principal focus. The intersection of these two reflected rays determines the position of the image of the point. Applying this construction to objects of sensible magnitude, it follows that the image of an object in a convex mirror is always *erect* and *diminished*.

107. If the mirror be *parabolic* instead of spherical, all parallel rays falling upon the mirror are collected to a point at its focus; conversely, a luminous point placed at the focus sends forth parallel rays: there is no aberration. If the mirror be *elliptical*, all rays emitted from one of the

REFRACTION OF LIGHT. 33

foci of the ellipsoid are collected together at the other. Parabolic reflectors are employed in light-houses, where it is an object to send a powerful beam, consisting of rays as nearly as possible parallel, far out to sea. In this case the centre of the flame is placed in the focus of the mirror; but, inasmuch as the flame is of sensible magnitude, and not a mere point, the rays of the reflected beam are not accurately parallel.

The Refraction of Light (Dioptrics).

108. We have hitherto confined our attention to the portion of a beam of light which rebounds from the reflecting surface. But, in general, a portion of the beam also *enters* the reflecting substance, being rapidly quenched when the substance is opaque (see note 11), and freely *transmitted* when the substance is transparent.

109. Thus in the case of water, mentioned in note 60, when the incidence is perpendicular all the rays are transmitted, save the 18 referred to as being reflected. That is to say, 982 out of every 1,000 rays enter the water and pass through it.

110. So likewise in the case of mercury, mentioned in the same note; 334 out of every 1,000 rays falling on the mercury at a perpendicular incidence, enter the metal and are quenched at a minute depth beneath its surface.

We have now to consider that portion of the luminous beam which enters the reflecting substance; taking, as an illustrative case, the passage from air into water.

111. If the beam fall upon the water as a perpendicular, it pursues a straight course through the water: if the incidence be oblique, the direction of the beam is changed at the point where it enters the water.

112. This bending of the beam is called *refraction*. Its amount is different in different substances.

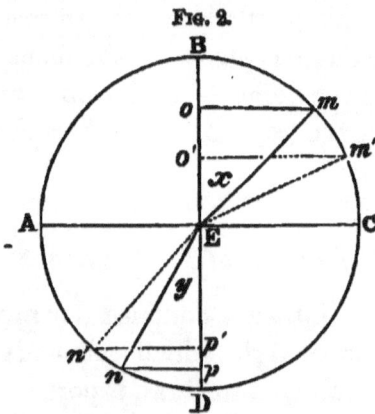

Fig. 2.

113. The refraction of light obeys a perfectly rigid law which must be clearly understood. Let A B C D, Fig. 2, be the section of a cylindrical vessel which is half filled with water, its surface being A C. E is the centre of the circular section of the cylinder, and B D is a perpendicular to the surface at E. Let the cylindrical envelope of the vessel be opaque, say of brass or tin, and let an aperture be imagined in it at B, through which a narrow light-beam passes to the point E. The beam will pursue a straight course to D without turning to the right or to the left.

114. Let the aperture be imagined at *m*, the beam striking the surface of the water at E *obliquely*. Its course on entering the liquid will be changed; it will pursue the track E *n*.

115. Draw the line *m o* perpendicular to B D, and also the line *n p* perpendicular to the same B D. It is always found that *m o* divided by *n p* is *a constant quantity*, no matter what may be the angle at which the ray enters the water.

116. The angle marked *x* above the surface is called the angle of incidence; the angle at *y* below the surface is called the angle of refraction; and if we regard the radius of the circle A B C D as unity or 1, the line *m o* will be the *sine* of the angle of incidence; while the line *n p* will be the *sine* of the angle of refraction.

117. Hence the ill-important optical law—*The sine of the angle of incidence divided by the sine of the angle of refraction is a constant quantity.* However these angles may vary in size, this bond of relationship is never severed. If one of them be lessened or augmented, the other must diminish or increase so as to obey this law. Thus if the incidence be along the dotted line *m'* E, the refraction will be along the line E *n'*, but the ratio of *m' o'* to *n' p'* will be precisely the same as that of *m o* to *n p*.

118. The constant quantity here referred to is called *the index of refraction.*

119. One word more is necessary to the full comprehension of the term *sine*, and of the experimental demonstration of the law of refraction. When one number is divided by another the quotient is called the *ratio* of the one number to the other. Thus 1 divided by 2 is $\frac{1}{2}$, and this is the ratio of 1 to 2. Thus also 2 divided by 1 is 2, and this is the ratio of 2 to 1. In like manner 12 divided by 3 is 4, and this is the ratio of 12 to 3. Conversely, 3 divided by 12 is $\frac{1}{4}$, and this is the ratio of 3 to 12.

120. In a right-angled triangle the ratio of any side to the hypothenuse is found by dividing that side by the hypothenuse. *This ratio is the sine of the angle opposite to the side,* however large or small the triangle may be. Thus in Fig. 2 the sine of the angle *x* in the right-angled triangle E *o m* is really the ratio of the line *o m* to the hypothenuse E *m*; it would be expressed in a fractional form thus, $\frac{o\ m}{E\ m}$. In like manner the sine of *y* is the ratio

of the line $n\,p$ to the hypothenuse $E\,n$, and would be expressed in a fractional form thus, $\dfrac{n\,p}{E\,n}$. These fractions are the sines of the respective angles, whatever be the length of the line $E\,m$ or $E\,n$. In the particular case above referred to, where these lines are considered as units, the fractions $\dfrac{m\,o}{1}$ and $\dfrac{n\,p}{1}$, or in other words $m\,o$ and $n\,p$, become, as stated, the sines of the respective angles. We are now prepared to understand a simple but rigid demonstration of the law of refraction.

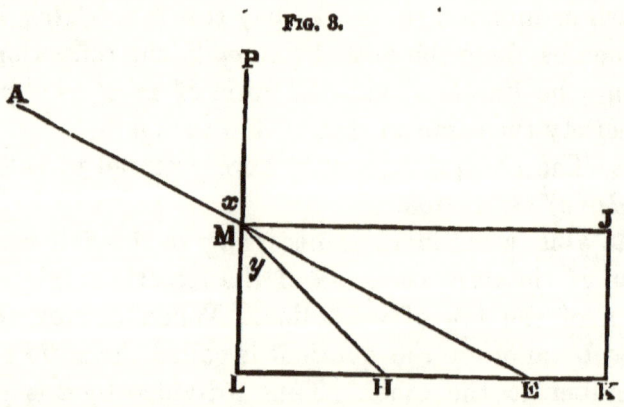

Fig. 3.

121. M L J K is a cell with parallel glass sides and one opaque end M L. The light of a candle placed at A falls into the vessel, the end M L casting a shadow which reaches to the point E. Fill the vessel with water—the shadow retreats to H through the refraction of the light at the point where it enters the water.

122. The angle enclosed between M E and M L is equal to the angle of incidence x, and, in accordance with the definition given in 120, $\dfrac{L\,E}{M\,E}$ is its sine; while $\dfrac{L\,H}{M\,H}$ is the sine of the angle of refraction y. All these lines can be

either measured or calculated. If they be thus determined, and if the division be actually made, it will always be found that the two quotients $\frac{LE}{ME}$ and $\frac{LH}{MH}$ stand in a constant ratio to each other, whatever the angle may be at which the light from A strikes the surface of the liquid. This ratio in the case of water is $\frac{4}{3}$, or, expressed in decimals, 1.333.*

123. When the light passes from air into water, the refracted ray is bent *toward* the perpendicular. This is generally, but not always, the case when the light passes from a rarer to a denser medium.

124. The principle of reversibility which runs through the whole of optics finds illustration here. When the ray passes from water to air it is bent *from* the perpendicular: it accurately reverses its course.

125. If instead of water we employed vinegar the ratio would be 1.344; with brandy it would be 1.360; with rectified spirit of wine 1.372; with oil of almonds or with olive oil 1.470; with spirit of turpentine 1.605; with oil of aniseed 1.538; with oil of bitter almonds 1.471; with bisulphide of carbon 1.678; with phosphorus 2.24.

126. These numbers express the indices of refraction of the various substances mentioned; all of them refract the light more powerfully than water, and it is worthy of remark that, all of them, except vinegar, are *combustible* substances.

127. It was the observation on the part of Newton, that, having regard to their density, "unctuous substances" as a general rule refracted light powerfully, coupled with the fact that the index of refraction of the diamond reached, according to his measurements, so high

* More accurately, 1.336.

a figure as 2.439, that caused him to foresee the possible combustible nature of the diamond. The bold prophecy of Newton* has been fulfilled, the combustion of a diamond being one of the commonest experiments of modern chemistry.

128. It is here worth noting that the refraction by spirit of turpentine is greater than that by water, though the density of the spirit is to that of the water as 874 is to 1,000. A ray passing obliquely from the spirit of turpentine into water is bent *from* the perpendicular, though it passes from a rarer to a denser medium; while a ray passing from water into the spirit of turpentine is bent *toward* the perpendicular, though it passes from a denser to a rarer medium. Hence the necessity for the words "not always" employed in 123.

129. If a ray of light pass through a refracting plate with parallel surfaces, or through any number of plates with parallel surfaces, on regaining the medium from which it started, its original direction is restored. This follows from the principle of reversibility already referred to.

130. In passing through a refracting body, or through any number of refracting bodies, the light accomplishes its transit in the *minimum of time*. That is to say, given the velocity of light in the various media, the path chosen by the ray, or, in other words, the path which its refraction imposes upon the ray, enables it to perform its journey in the most rapid manner possible.

131. Refraction always causes water to appear shallower, or a transparent plate of any kind thinner, than it

---

* " Car ce grand homme, qui mettait la plus grande sévérité dans ses expériences, et la plus grande réserve dans ses conjectures, n'hésitait jamais à suivre les conséquences d'une vérité aussi loin qu'elle pouvait le conduire."—BIOT

really is. The lifting up of the lower surface of a glass cube, through this cause, is very remarkable.

132. To understand why the water appears shallower, fix your attention on a point at its bottom, and suppose the line of vision from that point to the eye to be perpendicular to the surface of the water. Of all rays issuing from the point, the perpendicular one alone reaches the eye without refraction. Those close to the perpendicular, on emerging from the water, have their divergence augmented by refraction. Producing these divergent rays backward, they intersect at a point above the real bottom, and at this point the bottom will be seen.

133. The apparent shallowness is augmented by looking obliquely into the water.

134. In consequence of this apparent rise of the bottom, a straight stick thrust into the water is bent at the surface *from* the perpendicular.

Note the difference between the deportment of the stick and of a luminous beam. The beam on entering the water is bent *toward* the perpendicular.

135. This apparent lifting of the bottom when water is poured into a basin brings into sight an object at the bottom of the basin which is unseen when the basin is empty.

Opacity of Transparent Mixtures.

136. Reflection always accompanies refraction; and if one of these disappear, the other will disappear also. A solid body immersed in a liquid having the same refractive index as the solid, vanishes; it is no more seen than a portion of the liquid itself of the same size would be seen.

137. But in the passage from one medium to another of a different refractive index, light is always reflected; and this reflection may be so often repeated as to render the mixture of two transparent substances practically im-

pervious to light. It is the frequency of the reflections at the limiting surfaces of air and water that renders *foam* opaque. The blackest clouds owe their gloom to this repeated reflection, which diminishes their *transmitted* light. Hence also their whiteness by *reflected* light. To a similar cause is due the whiteness and imperviousness of common salt, and of transparent bodies generally when crushed to powder. The individual particles transmit light freely; but the reflections at their surfaces are so numerous that the light is wasted in echoes before it can reach to any depth in the powder.

138. The whiteness and opacity of writing-paper are due mainly to the same cause. It is a web of transparent fibres, not in optical contact, which intercept the light by repeatedly reflecting it.

139. But if the interstices of the fibres be filled by a body of the same refractive index as the fibres themselves, the reflection at their limiting surfaces is destroyed, and the paper is rendered transparent. This is the philosophy of the tracing-paper used by engineers. It is saturated with some kind of oil, the lines of maps and drawings being easily copied *through it* afterward. Water augments the transparency of paper, as it darkens a white towel; but its refractive index is too low to confer on either any high degree of transparency. It, however, renders certain minerals, which are opaque when dry, translucent.

140. The higher the refractive index the more copious is the reflection. The refractive index of water, for example, is 1.336; that of glass is 1.5. Hence the different quantities of light reflected by water and glass at a perpendicular incidence, as mentioned in note 60. It is its enormous refractive strength that confers such brilliancy upon the diamond.

Total Reflection.

Read notes 123 and 124; then continue here.

141. When the angle of incidence from air into water is nearly 90°, that is to say, when the ray before entering the water just grazes its surface, the angle of refraction is 48½°. Conversely, when a ray passing from water into air strikes the surface at an angle of 48½°, it will, on its emergence, just graze the surface of the water.

142. If the angle which the ray in water encloses with the perpendicular to the surface be greater than 48½°, the ray will not quit the water at all: it will be *totally reflected* at the surface.

143. The angle which marks the limit where total reflection begins is called the *limiting angle* of the medium. For water this angle is 48° 27', for flint glass it is 38° 41', while for diamond it is 23° 42'.

144. Realize clearly that a bundle of light rays filling an angular space of 90° before they enter the water, are squeezed into an angular space of 48° 27' within the water, and that in the case of diamond the condensation is from 90° to 23° 42'.

145. To an eye in still water its margin must appear *lifted up*. A fish, for example, sees objects, as it were, through a circular aperture of about 97° (twice 47° 27') in diameter overhead. All objects down to the horizon will be visible in this space, and those near the horizon will be much distorted and contracted in dimensions, especially in height. Beyond the limits of this circle will be seen the bottom of the water totally reflected, and therefore depicted as vividly as if seen by direct vision.*

146. A similar effect, exerted by the atmosphere (when

---

* Sir John Herschel.

no clouds cross the orbs), gives the sun and moon at rising and setting a slightly flattened appearance.

147. *Experimental Illustrations.*—Place a shilling in a drinking-glass; cover it with water to about the depth of an inch, and tilt the glass so as to obtain the necessary obliquity of incidence at the surface. Looking upward toward the surface, the image of the shilling will be seen shining there, and, as the reflection is total, the image will be as bright as the shilling itself. A spoon suitably dipped into the glass also yields an image due to total reflection.

148. Thrust the closed end of an empty test-tube into a glass of water. Incline the tube, until the horizontal light falling upon it shall be totally reflected upward. When looked down upon, the tube appears shining like burnished silver. Pour a little water into the tube: as the liquid rises, it abolishes total reflection, and with it the lustre, leaving a gradually diminishing lustrous zone, which disappears wholly when the level of the water within rises to, or above, that of the water without. A tube of any kind stopped water-tight will answer for this experiment, which is both beautiful and instructive.

149. If a ray of light fall as a perpendicular on the side of a right-angled isosceles glass prism, it will enter the glass and strike the hypothenuse at an angle of 45°. This exceeds the limiting angle of glass; the ray will therefore be totally reflected; and, in accordance with the law mentioned in note 54, the direct and reflected rays will be at right angles to each other. When such a change of direction is required in optical instruments, a right-angled isosceles prism is usually employed.

150. When the ray enters the prism parallel to the hypothenuse, it will be refracted, and will strike the hypothenuse at an angle greater than the limiting angle. It

will, therefore, be totally reflected. If the object, instead of being a point, be of sensible magnitude, the rays from its extremities will *cross each other* within the prism, and hence the object will appear *inverted* when looked at through the prism. Dove has applied the "reversion prism" to render erect the inverted images of the astronomical telescope.

151. The mirage of the desert and various other phantasmal appearances in the atmosphere are, in part, due to total reflection. When the sun heats an expanse of sand, the layer of air in contact with the sand becomes lighter than the superincumbent air. The rays from a distant object, a tree for example, striking very obliquely upon the upper surface of this layer, may be totally reflected, thus showing images similar to those produced by a surface of water. The thirsty soldiers of the French army were tantalized by such appearances in Egypt.

152. Gases, like liquids and solids, can refract and reflect light; but, in consequence of the lowness of their refractive indices, both reflection and refraction are feeble. Still, atmospheric refraction has to be taken into account by the astronomer, and by those engaged in trigonometrical surveys. The refraction of the atmosphere causes the sun to be seen before it actually rises, and after it actually sets.

153. The quivering of objects seen through air rising over a heated surface is due to irregular refraction, which incessantly shifts the directions of the rays of light. In the air this shifting of the rays is never entirely absent, and it is often a source of grievous annoyance to the astronomer who needs a homogeneous atmosphere.

154. The flame of a candle or of a gas-lamp, and the column of heated air above the flame; the air rising from

a red-hot iron; the pouring of a heavy gas, such as carbonic acid, downward into air; and the issue of a lighter one, such as hydrogen, upward — may all be made to reveal themselves by their action upon a sufficiently intense light. The transparent gases interposed between such a light and a white screen are seen to rise like smoke upon the screen through the effects of refraction.

Lenses.

155. A lens in optics is a portion of a refracting substance, such as glass, which is bounded by curved surfaces. If the surface be spherical, the lens is called a spherical lens.

156. Lenses divide themselves into two classes, one of which renders parallel rays convergent, the other of which renders such rays divergent. Each class comprises three kinds of lenses, which are named as follows:

Converging Lenses.

1. Double convex, with both surfaces convex.
2. Plano-convex, with one surface plane and the other convex.
3. Concavo-convex (Meniscus), with a concave and a convex surface, the convex surface being the most strongly curved.

Diverging Lenses.

1. Double concave, with both surfaces concave.
2. Plano-concave, with one surface plane and the other concave.
3. Convexo-concave, with a convex and a concave surface, the concave surface being the most strongly curved.

157. A straight line drawn through the centre of the lens, and perpendicular to its two convex surfaces, is the principal axis of the lens.

158. A luminous beam falling on a convex lens parallel to the axis, has its constituent rays brought to intersection at a point in the axis behind the lens. This point is the principal focus of the lens. As before, the principal focus is the focus of parallel rays.

159. The rays from a luminous point placed beyond the focus intersect at the opposite side of the lens, an image of the point being formed at the place of intersection. As the point approaches the principal focus its image retreats from it, and when the point actually reaches the principal focus, its image is at an infinite distance.

160. If the principal focus be passed, and the point come between that focus and the lens, the rays after passing through the lens will be still divergent. Producing them backward, they will intersect on that side of the lens on which stands the luminous point. The focus here is *virtual*. A body of sensible magnitude placed between the focus and the lens would have a virtual image.

161. When an object of sensible dimensions is placed anywhere beyond the principal focus, a real image of the object will be formed in the air behind the lens. The image may be either greater or less than the object in size, but the image will always be *inverted*.

162. The positions of the image and the object are, as before, convertible.

163. In the case of concave lenses the images are always virtual.

164. A spherical lens is incompetent to bring all the rays that fall upon it to the same focus. The rays which pass through the lens near its circumference are more

refracted than those which pass through the central portions, and they intersect earlier. Where perfect definition is required it is therefore usual, though at the expense of illumination, to make use of the central rays only.

165. This difference of focal distance between the central and circumferential rays is called the *spherical aberration* of the lens. A lens so curved as to bring all rays to the same focus is called *aplanatic;* a spherical lens cannot be rendered aplanatic.

166. As in the case of spherical mirrors, spherical lenses have their caustic curves and surfaces (diacaustics) formed by the intersection of the refracted rays.

Vision and the Eye.

167. The human eye is a compound lens, consisting of three principal parts: the *aqueous humor*, the *crystalline lens*, and the *vitreous humor*.

168. The aqueous humor is held in front of the eye by the *cornea*, a transparent, horny capsule, resembling a watch-glass in shape. Behind the aqueous humor, and immediately in front of the crystalline lens, is the *iris*, which surrounds the *pupil.* Then follow the lens and the vitreous humor, which last constitutes the main body of the eye. The average diameter of the human eye is 10.9 lines.*

169. When the optic nerve enters the eye from behind, it divides into a series of filaments, which are woven together to form the *retina*, a delicate net-work spread as a screen at the back of the eye. The retina rests upon a black pigment, which reduces to a minimum all internal reflection.

170. By means of the iris the size of the pupil may be caused to vary within certain limits. When the light is

* A line is $\frac{1}{12}$th of an inch.

feeble the pupil expands, when it is intense the pupil contracts; thus the quantity of light admitted into the eye is, to some extent, regulated. The pupil also diminishes when the eye is fixed upon a near object, and expands when it is fixed upon a distant one.

171. The pupil appears black; partly because of the internal black coating, but mainly for another reason. Could we illuminate the retina, and see at the same time the illuminated spot, the pupil would appear bright. But the principle of reversibility, so often spoken of, comes into play here. The light of the illuminated spot in returning outward retraces its steps, and finally falls upon the source of illumination. Hence, to receive the returning rays, the observer's eye ought to be placed between the source and the retina. But in this position it would cut off the illumination. If the light be thrown into the eye by a mirror pierced with a small orifice, or with a small portion of the silvering removed, then the eye of the observer placed behind the mirror, and looking through the orifice, may see the illuminated retina. The pupil under these circumstances glows like a live coal. This is the principle of the *Ophthalmoscope* (Augenspiegel, Helmholtz), an instrument by which the interior of the eye may be scanned, and its condition in health or disease noted.

172. In the case of albinos, or of white rabbits, the black pigment is absent, and the pupil is seen red by light which passes through the *sclerotica*, or white of the eye. When this light is cut off, the pupil of an albino appears black. In some animals the black pigment is displaced by a reflecting membrane, the *tapetum*. It is the light reflected from the tapetum which causes a cat's eye to shine in partial darkness. The light in this case is not internal, for when the darkness is *total* the cat's eyes do not shine.

173. In the camera obscura of the photographer the images of external objects formed by a convex lens are received upon a plate of ground glass, the lens being pushed in or out until the image upon the glass is sharply defined.

174. The eye is a camera obscura, with its refracting lenses, the retina playing the part of the plate of ground glass in the ordinary camera. For perfectly distinct vision it is necessary that the image upon the retina should be perfectly defined; in other words, that the rays from every point of the object looked at should be converged to a *point* upon the retina.

175. The image upon the retina is *inverted*.

Adjustment of the Eye: Use of Spectacles.

176. If the letters of a book held at some distance from the eye be looked at through a gauze veil placed nearer the eye, it will be found that when the letters are seen distinctly, the veil is seen indistinctly; conversely, if the veil be seen distinctly, the letters will be seen indistinctly. This demonstrates that the images of objects at different distances from the eye cannot be defined *at the same time* upon the retina.

177. Were the eye a rigid mass, like a glass lens, incapable of change of form, distinct vision would only be possible at one particular distance. We know, however, that the eye possesses a power of adjustment for different distances. This adjustment is effected, not by pushing the front of the eye backward or forward, but by changing the curvature of the crystalline lens.

178. The image of a candle reflected from the forward or backward surface of the lens is seen to diminish when the eye changes from distant to near vision, thus proving

the curvature of the lens to be greater for near than for distant vision.

179. The principal refraction endured by rays of light in crossing the eye occurs at the surface of the cornea, where the passage is from air to a much denser medium. The refraction at the cornea alone would cause the rays to intersect at a point nearly half an inch behind the retina. The convergence is augmented by the crystalline lens, which brings the point of intersection forward to the retina itself.

180. A line drawn through the centre of the cornea and the centre of the whole eye to the retina is called the axis of the eye. The length of the axis, even in youth, is sometimes too small; in other words, the retina is sometimes too near the cornea; so that the refracting part of the organ is unable to converge the rays from a luminous point so as to bring them to a point upon the retina. In old age also the refracting surfaces of the eye are slightly flattened, and thus rendered incompetent to refract the rays sufficiently. In both these cases the image would be formed *behind* the retina, instead of *upon* it, and hence the vision is indistinct.

181. The defect is remedied by holding the object at a distance from the eye, so as to lessen the divergence of its rays, or by placing in front of the eye a convex lens, which helps the eye to produce the necessary convergence. This is the use of spectacles.

182. The eye is also sometimes too long in the direction of the axis, or the curvature of the refracting surfaces may be too great. In either case the rays entering the pupil are converged so as to intersect before reaching the retina. This defect is remedied either by holding the object very close to the eye, so as to augment the divergence of its rays, thus throwing back the point of intersection; or

by placing in front of the eye a concave lens, which produces the necessary divergence.

183. The eye is not adjusted at the same time for equally-distant horizontal and vertical objects. The distance of distinct vision is greater for horizontal lines than for vertical ones. Draw with ink two lines at right angles to each other, the one vertical, the other horizontal: see one of them distinctly black and sharp; the other appears indistinct, as if drawn in lighter ink. Adjust the eye for this latter line, the former will then appear indistinct. This difference in the curvature of the eye in two directions may sometimes become so great as to render the application of cylindrical lenses necessary for its correction.

The Punctum Cœcum.

184. The spot where the optic nerve enters the eye, and from which it ramifies to form the net-work of the retina, is insensible to the action of light. An object whose image falls upon that spot is not seen. The image of a clock-face, of a human head, of the moon, may be caused to fall upon this "blind spot," and when this is the case the object is not visible.

185. To illustrate this point, proceed thus: Lay two white wafers on black paper, or two black ones on white paper, with an interval of 3 inches between them. Bring the right eye at a height of 10 or 11 inches exactly over the left-hand wafer, so that the line joining the two eyes shall be parallel to the line joining the two wafers. Closing the left eye, and looking steadily with the right at the left-hand wafer, the right-hand one ceases to be visible. In this position the image falls upon the "blind spot" of the right eye. If the eye be turned in the least degree to the right or left, or if the distance between it and the paper

be augmented or diminished, the wafer is immediately seen. Preserving these proportions as to size and distance, objects of far greater dimensions than the wafer may have their images thrown upon the blind spot, and be obliterated.

Persistence of Impressions.

186. An impression of light once made upon the retina does not subside instantaneously. An electric spark is sensibly instantaneous; but the impression it makes upon the eye remains for some time after the spark has passed away. This interval of persistence varies with different persons, and amounts to a sensible fraction of a second.

187. If, therefore, a succession of sparks follow each other at intervals less than the time which the impression endures, the separate impressions will unite to form *a continuous* light. If a luminous point be caused to describe a circle in less than this interval, the circle will appear as a continuous closed curve. From this cause, also, the spokes of a rapidly-rotating wheel blend together to a shadowy surface. Wheatstone's Photometer is based on this persistence. It also explains the action of those instruments in which a series of objects in different positions being brought in rapid succession before the eye, the impression of *motion* is produced.

188. A jet of water descending from an orifice in the bottom of a vessel exhibits two distinct parts: a tranquil pellucid portion near the orifice, and a turbid or untranquil portion lower down. Both parts of the jet appear equally continuous. But when the jet in a dark room is illuminated by an electric spark, all the turbid portion reveals itself as a string of separate drops standing perfectly still. It is their quick succession that produces the impression of continuity. The most rapid cannon-ball, illuminated by

a flash of lightning, would be seen for the fraction of a second perfectly motionless in the air.

189. The eye is by no means a perfect optical instrument. It suffers from spherical aberration; a scattered luminosity, more or less strong, always surrounding the defined images of luminous objects upon the retina. By this luminosity the image of the object is sensibly increased in size; but with ordinary illumination the scattered light is too feeble to be noticed. When, however, bodies are intensely illuminated, more especially when the bodies are small, so that a slight extension of their images upon the retina becomes noticeable, such bodies appear larger than they really are. Thus, a platinum-wire raised to whiteness by a voltaic current has its apparent diameter enormously increased. Thus also the crescent moon seems to belong to a larger sphere than the dimmer mass of the satellite which it partially clasps. Thus also, at considerable distances, the parallel flashes sent from a number of separate lamps and reflectors in a light-house encroach upon each other, and blend together to a single flash. The white-hot particles of carbon in a flame describe *lines* of light, because of their rapid upward motion. These lines are *widened* to the eye; and thus a far greater apparent *solidity* is imparted to the flame than in reality belongs to it.

189*a*. This augmentation of the true size of the optical image is called *Irradiation*.

Bodies seen within the Eye.

190. Almost every eye contains bodies more or less opaque distributed through its humors. The so-called *muscæ volitantes* are of this character; so are the black dots, snake-like lines, beads, and rings, which are strikingly visible in many eyes. Were the area of the pupil contracted to a point, such bodies might produce considerable

annoyance; but because of the width of the pupil the shadows which these small bodies would otherwise cast upon the retina are practically obliterated, except when they are very close to the back of the eye.* It is only necessary to look at the firmament through a pinhole to give these shadows greater definition upon the retina.

191. The veins and arteries of the retina itself also cast their shadows upon its posterior surface; but the shaded spaces soon become so sensitive to light as to compensate for the defect of light falling upon them. Hence under ordinary circumstances the shadows are not seen. But if the shadows be transported to a less sensitive portion of the retina, the image of the vessels becomes distinctly visible.

192. The best mode of obtaining the transference of the shadow is to concentrate in a dark room, by means of a pocket lens of short focus, a small image of the sun or of the electric light upon the *white* of the eye. Care must be taken not to send the beam through the pupil. When the small lens is caused to move to and fro, the shadows are caused to travel over different portions of the retina, and a perfectly defined image of the veins and arteries is seen projected in the darkness in front of the eye.

193. Looking into a dark space, and moving a candle at the same time to and fro beside the eye, so that the rays enter the pupil very obliquely, the shadow of the retinal vessels is also obtained. In some eyes the suddenness and vigor with which the spectral image displays itself are extraordinary; others find it difficult to obtain the effect.

194. Finally, a delicate image of the vessels may be obtained by looking through a pinhole at the bright sky, and moving the aperture to and fro.

* See Notes 18 and 19.

The Stereoscope.

195. Look with one eye at the edge of the hand, so that the finger nearest the eye shall cover all the others. Then open the second eye; by it the other fingers will be seen foreshortened. *The images of the hand therefore within the two eyes are different.*

196. These two images are projected on the two retinæ; if by any means we could combine two drawings, executed on a flat surface, so as to produce within the two eyes two pictures similar to the two images of the solid hand, we should obtain the impression of *solidity*. This is done by the stereoscope.

197. The first form of this instrument was invented by Sir Charles Wheatstone. He took drawings of solid objects as seen by the two eyes, and looked at the images of these drawings in two plane mirrors. Each eye looked at the image which belonged to it, and the mirrors were so arranged that the images overlapped, thus appearing to come from the same object. By this combination of its two plane projections, the object sketched was caused to start forth as a solid.

198. In looking at and combining two such drawings, the eyes receive the same impression, and go through the same process as when they look at the real object. We see only one point of an object distinctly at a time. If the different points of an object be at different distances from the eyes, to see the near points distinctly requires a greater convergence of the axes of the eyes than to see the distant ones. Now, besides the identity of the retinal images of the stereoscopic drawings with those of the real object, the eyes, in order to cause the corresponding pairs of points of the two drawings to coalesce, have to go through the same

THE STEREOSCOPE. 55

variations of convergence that are necessary to see distinctly the various points of the actual object. Hence the impression of solidity produced by the combination of such drawings.

199. Measure the distance between *two pairs* of points, which when combined by the stereoscope present two *single points* at different distances from the eye. The distance between the one pair will be greater than that between the other pair. Different degrees of convergence are therefore necessary on the part of the eye to combine the two pairs of points. It is to be noted that the coalescence produced in the stereoscope at any particular moment is only *partial*. If one pair of corresponding points be seen singly, the others must appear double. This is also the case when an actual solid is looked at with both eyes; of those points of it which are at different distances from the eyes one only is seen singly at a time.

200. The impression of solidity may be produced in an exceedingly striking manner without any stereoscope at all. Most easily, thus: Take two drawings—projections, as they are called—of the frustum of a cone; the one as it is seen by the right eye, the other as it is seen by the left. Holding them at some distance from the eyes, let the left-hand drawing be looked at by the right eye, and the right-hand drawing by the left. The lines of vision of the two eyes here cross each other; and it is easy, after a few trials with a pencil-point placed in front of the eyes, to make two corresponding points of the drawings coincide. The moment they coincide, the combined drawings start forth as a single solid, suspended in the air at the place of intersection of the lines of vision. It depends upon the character of the drawings whether the inside of the frustum is seen, or the outside, whether its base or its top seems nearest to the eye. For this experiment the drawings are best made

in simple outline, and they may be immensely larger than ordinary stereoscopic drawings.

Take notice that here also the different pairs of the corresponding points are at different distances apart. Two corresponding points, for example, of the top of the frustum will not be the same distance asunder as two points of the base.

201. Wheatstone's first instrument is called the Reflecting Stereoscope; but the methods of causing drawings to coalesce so as to produce stereoscope effects are almost numberless. The instrument most used by the public is the Lenticular Stereoscope of Sir David Brewster. In it the two projections are combined by means of two half lenses with their edges turned inward. The lenticular stereoscope also magnifies.*

202. It has been stated in note 198 that for the distinct vision of a near point a greater convergence of the lines of vision of the two eyes is necessary than that of a distant one. By an instrument in which two rectangular prisms are employed,† the rays from two points may be caused to *cross* before they enter the eyes, the convergence being thus rendered greater for the distant point than for the near one. The consequence of this is, that the near point appears distant, and the distant point near. This is the principle of Wheatstone's *pseudoscope*. By this instrument convex surfaces are rendered concave, and concave surfaces convex. The inside of a hat or teacup may be thus converted into its outside, while a globe may be seen as a concave spherical surface.

* Fuller and clearer information regarding the stereoscope will be found in the *Journal of the Photographic Society*, vol. iii. pp. 96, 116, and 167.

† See Note 150.

Nature of Light; Physical Theory of Reflection and Refraction.

It is now time to redeem to some extent the promise of our first note, that the "something" which excites the sensation of light should be considered more closely subsequently.

203. Every sensation corresponds to a motion excited in our nerves. In the sense of touch, the nerves are moved by the contact of the body felt; in the sense of smell, they are stirred by the infinitesimal particles of the odorous body; in the sense of hearing, they are shaken by the vibrations of the air.

Theory of Emission.

204. Newton supposed light to consist of small particles shot out with inconceivable rapidity by luminous bodies, and fine enough to pass through the pores of transparent media. Crossing the humors of the eye, and striking the optic nerve behind the eye, these particles were supposed to excite vision.

205. This is the *Emission Theory* or *Corpuscular Theory* of Light.

206. Considering the enormous velocity of light, the particles, if they exist, must be inconceivably small; for if of any conceivable weight, they would infallibly destroy so delicate an organ as the eye. A bit of ordinary matter, one grain in weight, and moving with the velocity of light, would possess the momentum of a cannon-ball 150 lbs. weight, moving with a velocity of 1,000 feet a second.

207. Millions of these light particles, supposing them to exist, concentrated by lenses and mirrors, have been shot against a balance suspended by a single spider's thread; this thread, though twisted 18,000 times, showed

no tendency to untwist itself; it was therefore devoid of torsion. But no motion due to the impact of the particles was even in this case observed.*

208. If light consists of minute particles, they must be shot out with the same velocity by all celestial bodies. This seems exceedingly unlikely, when the different gravitating forces of such different masses are taken into account. By the attractions of such diverse masses, the particles would in all probability be pulled back with different degrees of force.

209. If, for example, a fixed star of the sun's density possessed 250 times the sun's diameter, its attraction, supposing light to be acted on like ordinary matter, would be sufficient to finally stop the particles of light issuing from it. Smaller masses would exert corresponding degrees of retardation; and hence the light emitted by different bodies would move with different velocities. That such is not the case—that light moves with the same velocity whatever be its source—renders it probable that it does *not* consist of particles thus darted forth.

But a more definite and formidable objection to the Emission Theory may be stated after we have made ourselves acquainted with the account it rendered of the phenomena of reflection and refraction.

210. In direct reflection, according to the emission theory, the light particles are first of all stopped in their course by a repellent force exerted by the reflecting body, and then driven in the contrary direction by the same force.

211. This repulsion is, however, *selective*. The reflecting substance singles out one portion of the group of particles composing a luminous beam and drives them back;

* Bennett, *Phil. Trans.*, 1792.

but it attracts the remaining particles of the group and transmits them.

212. When a light particle approaches a refractive surface obliquely, if the particle be an attracted one, it is drawn toward the surface, as an ordinary projectile is drawn toward the earth. Refraction is thus accounted for. Like the projectile, too, the velocity of the light particle is *augmented* during its deflection; it enters the refracting medium with this increased velocity, and once within the medium, the attractions before and behind the particle neutralizing each other, the increased velocity is maintained.

213. Thus, it is an unavoidable consequence of the theory of Newton, that the bending of a ray of light toward the perpendicular is accompanied by an augmentation of velocity—that light in water moves more rapidly than in air, in glass more rapidly than in water, in diamond more rapidly than in glass. In short, that the higher the refractive index, the greater the velocity of the light.

214. A decisive test of the emission theory was thus suggested, and under that test the theory has broken down. For it has been demonstrated, by the most rigid experiments, that the velocity of light *diminishes* as the index of refraction increases. The theory, however, had yielded to the assaults made upon it long before this particular experiment was made.

Theory of Undulation.

215. The Emission Theory was first opposed by the celebrated astronomer Huyghens, and the no less celebrated mathematician Euler, both of whom held that light, like sound, was a product of *wave-motion.* Laplace, Malus, Biot, and Brewster, supported Newton, and the emission theory maintained its ground until it was finally over-

thrown by the labors of Thomas Young* and Augustin Fresnel.

216. These two eminent philosophers, while adducing whole classes of facts inexplicable by the emission theory, succeeded in establishing the most complete parallelism between optical phenomena and those of wave-motion. The justification of a theory consists in its exclusive competence to account for phenomena. On such a basis the *Wave Theory*, or the *Undulatory Theory* of light, now rests, and every day's experience only makes its foundations more secure. This theory must for the future occupy much of our attention.

217. In the case of sound, the velocity depends upon the relation of elasticity to density in the body which transmits the sound. The greater the elasticity the greater is the velocity, and the less the density the greater is the velocity. To account for the enormous velocity of propagation in the case of light, the substance which transmits it is assumed to be of both extreme elasticity and of ex-

---

* Dr. Young was appointed Professor of Natural Philosophy in the Royal Institution, August 3, 1801. From a marble slab in the village church of Farnborough, near Bromley, Kent, I copied, on the 11th of April, the following inscription:

"Near this place are deposited the remains of THOMAS YOUNG, M. D., Fellow and Foreign Secretary of the Royal Society, Member of the National Institute of France. A man alike eminent in almost every department of human learning, whose many discoveries enlarged the bounds of Natural Science, and who first penetrated the obscurity which had veiled for ages the Hieroglyphics of Egypt.

"Endeared to his friends by his domestic virtues, Honored by the world for his unrivalled acquirements, He died in the hope of the resurrection of the just.

"Born at Milverton, in Somersetshire, June 13, 1773.
"Died in Park Square, London, May 29, 1829,
"In the 56th year of his age."

THEORY OF UNDULATION.   61

treme tenuity. This substance is called the *Luminiferous ether*.

218. It fills space; it surrounds the atoms of bodies; it extends, without solution of continuity, through the humors of the eye. The molecules of luminous bodies are in a state of vibration. The vibrations are taken up by the ether, and transmitted through it in waves. These waves impinging on the retina excite the sensation of light.

219. In the case of sound, the air-particles oscillate to and fro in the direction *in* which the sound is transmitted; in the case of light, the ether particles oscillate to and fro *across* the direction in which the light is propagated. In scientific language the vibrations of sound are *longitudinal*, while the vibrations of light are *transversal*. In fact, the mechanical properties of the ether are rather those of a solid than of an air.

220. The *intensity* of the light depends on the distance to which the ether particles move to and fro. This distance is called the *amplitude* of the vibration. The intensity of light is proportional to the *square* of the amplitude; it is also proportional to the square of the maximum velocity of the vibrating particle.

221. The amplitude of the vibrations diminishes simply as the distance increases; consequently the intensity, which is expressed by the square of the amplitude, must diminish inversely as the square of the distance. This, in the language of the wave theory, is the law of inverse squares.

222. The reflection of ether waves obeys the law established in the case of light. The angle of incidence is demonstrably equal to the angle of reflection.

223. To account for refraction, let us for the sake of simplicity take a portion of a circular wave emitted by the

sun or some other distant body. A short portion of such a wave would be *straight*. Suppose it to impinge from air upon a plate of glass, the wave being in the first instance *parallel* to the surface of the glass. Such a wave would go through the glass without change of direction.

224. But as the velocity in glass is less than the velocity in air, the wave would be *retarded* on passing into the denser medium.

225. But suppose the wave, before impact, to be *oblique* to the surface of the glass; that end of the wave which first reaches the glass will be first retarded, the other portions being held back in succession. This retardation of one end of the wave causes it to swing round; so that when the wave has fully entered the glass its course is oblique to its first direction. It is *refracted*.

226. If the glass into which the wave enters be a plate with parallel surfaces, the portion of the wave which reached the upper surface *first*, and was first retarded, will also reach its under surface first, and escape earliest from the retarding medium. This produces a second swinging round of the wave, by which its original direction is restored. In this simple way the Wave Theory accounts for Refraction.

227. The convergence or divergence of beams of light by lenses is immediately deduced from the fact that the different points of the ether wave reach the lens, and are retarded by the lens in succession.

228. The density of the ether is greater in liquids and solids than in gases, and greater in gases than in vacuo. A compressing force seems to be exerted on the ether by the molecules of these bodies. Now if the elasticity of the ether increased in the same proportion as its density, the one would neutralize the other, and we should have no retardation of the velocity of light. The diminished ve-

locity in highly-refracting bodies is accounted for by assuming that in such bodies the elasticity *in relation to the density* is less than in vacuo. The observed phenomena immediately flow from this assumption.

229. The case is precisely similar to that of sound in a gas or vapor which does not obey the law of Mariotte. The elasticity of such a gas or vapor, when compressed, increases less rapidly than the density; hence the diminished velocity of the sound.

230. But we are able to give a more distinct statement as to the influence which a refracting body has upon the velocity of light. Regard the lines $o\,m$ and np in Fig. 2, Note 113. These two lines *represent the velocities of light* in the two media there considered; or, expressed more generally, the sine of the angle of incidence represents the velocity of light in the first medium, while the sine of the angle of refraction represents the velocity in the second. *The index of refraction then is nothing else than the ratio of the two velocities.* Thus in the case of water, where the index of refraction is $\frac{4}{3}$, the velocity in air is to its velocity in water as 4 is to 3. In glass also, where the index of refraction is $\frac{3}{2}$, the velocity in air is to the velocity in glass as 3 is to 2. In other words the velocity of light in air is $1\frac{1}{3}$ times its velocity in water, and $1\frac{1}{2}$ times its velocity in glass. The velocity of light in air is about $2\frac{1}{2}$ times its velocity in diamond, and nearly three times its velocity in chromate of lead, the most powerfully refracting substance hitherto discovered. Strictly speaking, the index of refraction refers to the passage of a ray of light, not from *air*, but from a vacuum,* into the refracting body. Dividing the velocity of light in vacuo by its velocity in the refracting substance, the quotient is the index of refraction of that substance.

* That is to say, a vacuum save as regards the ether itself

231. In the wave theory, the rays of light are perpendiculars to the waves of ether. Unlike the *wave*, the *ray* has no material existence; it is merely a direction.

Prisms.

232. It has been stated, in Note 129, that in the case of a plate of glass *with parallel surfaces*, the direction possessed by an oblique ray, prior to its meeting the glass, is restored when it quits the glass. This is not the case if the two surfaces at which the ray enters and emerges be not parallel.

233. When the ray passes through a wedge-shaped transparent substance, in a direction perpendicular to the edge of the wedge, it is *permanently* refracted. A body of this shape is called a *prism* in optics, and the angle enclosed by the two oblique sides of the wedge is called the *refracting angle*.

234. The larger the refracting angle the greater is the deflection of the ray from its original direction. But with the self-same prism the amount of the refraction varies with the direction pursued by the ray through the prism.

235. When that direction is such that the portion of the ray within the prism makes equal angles with the two sides of the prism, or, what is the same, with the ray before it reaches the prism and after it has quitted it, then the total refraction is a *minimum*. This is capable both of mathematical and experimental proof; and on this result is based a method of determining the index of refraction.

236. The final direction of a refracted ray being unaltered by its passage through glass plates with parallel surfaces, we may employ hollow vessels composed of such plates and filled with liquids, thus obtaining liquid prisms.

Prismatic Analysis of Light; Dispersion.

237. Newton first unravelled the solar light, proving it to be composed of an infinite number of rays of different degrees of refrangibility; when such light is sent through a prism, its constituent rays are drawn asunder. This act of drawing apart is called *dispersion*.

238. The waves of ether generated by luminous bodies are not all of the same length; some are longer than others. In refracting substances the short waves are *more retarded* than the longer ones; hence the short waves are more *refracted* than the long ones. This is the cause of dispersion.

239. The luminous image formed when a beam of white light is thus decomposed by a prism is called a *spectrum*. If the light employed be that of the sun, the image is called the solar spectrum.

240. The solar spectrum consists of a series of vivid colors, which, when reblended, produce the original white light. Commencing with that which is least refracted, we have the following order of colors in the solar spectrum: Red, Orange, Yellow, Green, Blue, Indigo, Violet.

241. *The Color of Light is determined solely by its Wave-length.*—The ether waves gradually diminish in length from the red to the violet. The length of a wave of red light is about $\frac{1}{39000}$th of an inch; that of the wave of violet light is about $\frac{1}{57000}$th of an inch. The waves which produce the other colors of the spectrum lie between these extremes.

242. The velocity of light being 192,000 miles in a second, if we multiply this number by 39,000 we obtain the number of waves of red light in 192,000 miles; the product is 474,439,680,000,000. *All these waves enter the eye in a second.* In the same interval 699,000,000,000,000

waves of violet light enter the eye. At this prodigious rate is the retina hit by the waves of light.

243. Color, in fact, is to light what *pitch* is to sound. The pitch of a note depends solely on the number of aerial waves which strike the ear in a second. The color of light depends on the number of ethereal waves which strike the eye in a second. Thus the sensation of red is produced by imparting to the optic nerve four hundred and seventy-four millions of millions of impulses, per second, while the sensation of violet is produced by imparting to the nerve six hundred and ninety-nine millions of millions of impulses per second. In the Emission Theory numbers not less immense occur, "nor is there any mode of conceiving the subject which does not call upon us to admit the exertion of mechanical forces which may well be termed infinite." *

Invisible Rays; Calorescence and Fluorescence.

244. The spectrum extends in both directions beyond its visible limits. Beyond the visible red we have rays which possess a high heating power, though incompetent to excite vision; beyond the violet we have a vast body of rays which, though feeble as regards heat, and powerless as regards light, are of the highest importance because of their capacity to produce chemical action.

245. In the case of the electric light, the energy of the non-luminous calorific rays emitted by the carbon points is about eight times that of all the other rays taken together. The dark calorific rays of the sun also probably exceed many times in power the luminous solar rays. It is possible to sift the solar or the electric beam so as to intercept the luminous rays, while the non-luminous rays are allowed free transmission.

* Sir John Herschel.

246. In this way perfectly dark foci may be obtained where combustible bodies may be burned, non-refractory metals fused, and refractory ones raised to the temperature of whiteness. The non-luminous calorific rays may be thus transformed into luminous ones, which yield all the colors of the spectrum. This passage, by the intervention of a refractory body, from the non-luminous to the luminous state, is called *Calorescence*.

247. So also as regards the ultra-violet rays; when they are permitted to fall upon certain substances—the disulphate of quinine for example—they render the substance luminous; *invisible rays are thereby rendered visible*. The change here receives the name of *Fluorescence*.

248. In calorescence the atoms of the refractory body are caused to vibrate more rapidly than the waves which fall upon them; the periods of the waves are quickened by their impact on the atoms. The refrangibility of the rays is, in fact, *exalted*. In fluorescence, on the contrary, the impact of the waves throws the molecules of the fluorescent body into vibrations of slower periods than those of the incident waves; the refrangibility of the rays is in fact *lowered*. Thus by exalting the refrangibility of the ultra-red, and by lowering the refrangibility of the ultra-violet rays, both classes of rays are rendered capable of exciting vision.

249. Though the term is by no means faultless, those rays, both ultra-red and ultra-violet, which are incompetent to excite vision, are called *invisible rays*. In strictness we cannot speak of rays being either visible or invisible; it is not the rays themselves but the objects they illuminate that become visible. "*Space*, though traversed by the rays from all suns and all stars, is itself unseen. Not even the ether which fills space, and whose motions are the light of the world, is itself visible."*

* "Proceedings of the Royal Institution," vol. v., p. 456.

Doctrine of Visual Periods.

250. A string tuned to a certain note resounds when that note is sounded. If you sing into an open piano, the string whose note is in unison with your voice will be thrown into sonorous vibration. If there be discord between the note and the string, there is no resonance, however powerful the note may be. A particular church-pane is sometimes broken by a particular organ-peal, through the coincidence of its period of vibration with that of the organ.

251. In this way it is conceivable that a feeble note, through the coincidence of its periods of vibration with those of a sounding body, may produce effects which a powerful note, because of its non-coincidence, is unable to produce.

252. This, which is a known phenomenon of sound, helps us to a conception of the deportment of the retina toward light. The retina, or rather the brain in which its fibres end, is, as it were, attuned to a certain range of vibrations, and it is dead to all vibrations which lie without that range, however powerful they may be.

253. The quantity of wave-motion sent to the eye at night, by a candle a mile distant, suffices to render the candle visible. Employing the powerful ultra-red rays of the sun, or of the electric light, it is demonstrable that ethereal waves possessing many millions of times the mechanical energy of those which produce the candle's light, may be caused to impinge upon the retina without exciting any sensation whatever. *The periods of succession* of the waves, rather than their *strength*, are here influential.

254. When in music two notes are separated from each other by an octave, the higher note vibrates with twice

the rapidity of the lower. In Note 241 the lengths of the wave of red light and of violet light are set down as $\frac{1}{39000}$ of an inch and $\frac{1}{57000}$ of an inch respectively; but these numbers refer to the *mean* red and the *mean* violet. The waves of the *extreme* violet are about half the length of those of the extreme red, and they strike the retina with double the rapidity of the red. While, therefore, the *musical scale*, or the range of the ear, is known to embrace nearly eleven octaves, the *optical scale*, or range of the eye, is comprised within a single octave.

Doctrine of Colors.

255. Natural bodies possess the power of extinguishing, or, as it is called, *absorbing* the light that enters them. This power of absorption is *selective*, and hence, for the most part, arise the phenomena of *color*.

256. When the light which enters a body is *wholly* absorbed the body is black; a body which absorbs all the waves equally, but not totally, is gray; while a body which absorbs the various waves unequally is *colored*. Color is due to the extinction of certain constituents of the white light within the body, the remaining constituents which return to the eye imparting to the body its color.

257. It is to be borne in mind that bodies of all colors, illuminated by white light, reflect white light *from their exterior surfaces*. It is the light which has plunged to a certain depth within the body, which has been *sifted* there by elective absorption, and then discharged from the body by interior reflection that, in general, gives the body its color.

258. A pure red glass interposed in the path of a beam decomposed by a prism, either before or after the act of decomposition, cuts off all the colors of the spectrum except the red. A glass of any other pure color similarly inter-

posed would cut off all the spectrum except that particular portion which gives the glass its color. It is, however, extremely difficult, if not impossible, to obtain pure pigments of any kind. Thus a yellow glass not only allows the yellow light of the spectrum to pass, but also a portion of the adjacent green and orange; while a blue glass not only allows the blue to pass, but also a portion of the adjacent green and indigo.

259. Hence, if a beam of white light be caused to pass through a yellow glass and a blue glass at the same time, the only transmissible color common to both is green. This explains why blue and yellow powders, when mixed together, produce green. The white light plunges into the powder to a certain depth, and is discharged by internal reflection, *minus* its yellow and its blue. The green alone remains.

260. The effect is quite different when, instead of mixing blue and yellow *pigments*, we mix blue and yellow *lights* together. Here the mixture is a pure *white*. Blue and yellow are complementary colors.

261. Any two colors whose mixture produces white are called *complementary colors*. In the spectrum we have the following pairs of such colors:

> Red and greenish Blue.
> Orange and cyanogen Blue.
> Yellow and indigo Blue.
> Greenish Yellow and Violet.

262. A body placed in a light which it is incompetent to transmit appears black, however intense may be the illumination. Thus, a stick of red sealing-wax, placed in the vivid green of the spectrum, is perfectly black. A bright-red solution similarly placed cannot be distinguished from black ink; and red cloth, on which the spectrum is

permitted to fall, shows its color vividly where the red light falls upon it, but appears black beyond this position.

263. We have thus far dealt with the *analysis* of white light. In reblending the constituent colors, so as to produce the original, we illustrate, by *synthesis*, the composition of white light.

264. Let the beam analyzed be a rectangular slice of light. By means of a cylindrical lens we can recombine the colors, and produce by their mixture the original white. It is also possible, by the combination of the colors of its spectrum, to build up a perfect image of the source of light. The persistence of impressions on the retina also offers a ready means of blending colors.

Chromatic Aberration.—Achromatism.

265. Owing to the different refrangibility of the different rays of the spectrum, it is impossible by a single spherical lens to bring them all to a focus at the same point. The blue rays, for example, being more refracted than the red will intersect sooner than the red.

266. Hence, when a divergent cone of white light is rendered convergent by a lens, the convergent beam, as far as the point of intersection of the rays, will be surrounded by a sheath of red; while beyond the focus the divergent cone will be surrounded by a sheath of blue. Hence, when the refracted rays fall upon a screen placed between the lens and the focus of blue rays, a white circle with a red border is obtained; while if the screen be placed beyond the focus of red rays, the white circle will have a blue border. It is impossible to produce a colorless image in these positions of the screen.

267. This lack of power on the part of a lens to bring its differently-colored constituents to a common focus, is called the *Chromatic aberration* of the lens.

268. Newton considered it impossible to get rid of chromatic aberration; for he supposed the dispersion of a prism or lens to be proportional to its refraction, and that if you destroyed the one you destroyed the other. This, however, was an error.

269. For two prisms producing the same mean refraction may produce very different degrees of dispersion. By diminishing the angle of the more highly-dispersive prism we can make its dispersion sensibly equal to that of the feebly dispersive one; and we can neutralize the colors of both prisms by placing them in opposition to each other, without neutralizing the refraction.

270. When, for example, a prism of water is opposed to a prism of flint-glass, after the dispersion of the water, which is small, has been destroyed, the beam is still refracted. If a prism of *crown-glass* be substituted for the prism of water, substantially the same effect is produced. The flint-glass is competent to neutralize the dispersion of the crown *before* it neutralizes the refraction.

271. What is here said of prisms applies equally to lenses. A convex crown-glass lens, opposed to a concave flint-glass lens, may have its dispersion destroyed, and still images may be formed by the combination of the two lenses, because of the *residual* refraction.

272. A combination of lenses wherein color is destroyed while a certain amount of refraction is preserved, is called an achromatic combination, or more briefly an *achromatic lens*.

273. The human eye is not achromatic. It suffers from chromatic aberration as well as from spherical aberration.

Subjective Colors.

274. By the action of light the optic nerve is rendered less sensitive. When we pass from bright daylight into a moderately-lighted room, the room appears dark.

SUBJECTIVE COLORS. 73

275. This is also true of individual colors; when light of any particular color falls upon the eye, the optic nerve is rendered less sensitive to that color. It is, in fact, partially blinded to its perception.

276. If the eyes be steadily fixed upon a red wafer placed on white paper, after a little time the wafer will be surrounded by a greenish rim, and if the wafer be moved away, the place on which it rested will appear green.

277. This is thus explained: the eye by looking at the wafer has its sensibility to red light diminished; hence, when the wafer is removed, the white light falling upon the spot of the retina on which the image of the wafer rested, will have its red constituent virtually removed, and will therefore appear of the complementary color. The first rim of green light observed is due to the extension of the red light of the wafer a little beyond its geometrical image on the retina, in consequence of the spherical aberration of the eye.

278. Colored shadows are reducible to the same cause. Let a strong red light, for example, fall upon a white screen. A body interposed between the light and the screen will cast a shadow, and if this shadow be moderately illuminated by a second white light it will appear green. If the original light be blue, the shadow will appear yellow; if the original light be green, the shadow will appear red. The reason is, that the eye in the first instance is partially blinded to the perception of the color cast upon the screen; hence the white light, which reaches the eye from the shadow, will have that color partially withdrawn, and the shadow will appear of the complementary color.

279. Colors of this kind are called *subjective colors;* they depend upon the condition of the eye, and do not express external facts of color.

Spectrum Analysis.

280. Metals and their compounds impart to flames peculiar colors, which are characteristic of the metals. Thus the almost lightless flame of a Bunsen's burner is rendered a brilliant yellow by the metal sodium, or by any volatilizible compound of that metal, such as chloride of sodium or common salt. The flame is rendered green by copper, purple by zinc, and red by strontian.

281. These colors are due to the *vapors* of the metals which are liberated in the flame.

282. When such incandescent metallic vapors are examined by the prism, it is found that instead of emitting rays which form a *continuous* spectrum, one color passing gradually into another, they emit distinct groups of rays of definite, but different refrangibilities. The spectrum corresponding to these rays is a series of colored bands, separated from each other by intervals of darkness. Such bands are characteristic of luminous gases of all kinds.

283. Thus the spectrum of incandescent sodium-vapor consists of a brilliant band on the confines of the orange and yellow; and the vapor is incompetent to shed forth any of the other light of the spectrum. When this band is more accurately analyzed it resolves itself into two distinct bands; greater delicacy of analysis resolves it into *a group* of bands with fine dark intervals between them. The spectrum of copper-vapor is signalized by a series of green bands, while the incandescent vapor of zinc produces brilliant bands of blue and red.

284. The light of the bands produced by metallic vapors is very intense, the whole of the light being concentrated into a few narrow strips, and escaping in a great measure the dilution due to dispersion.

285. These colored bands are perfectly characteristic

of the vapor; from their position and number the substance that produces them can be unerringly inferred.

286. If two or more metals be introduced into the flame at the same time, prismatic analysis reveals the bands of each metal as if the others were not there. This is also true when a mineral containing several metals is introduced into the flame. The constituent metals of the mineral will give each its characteristic bands.

287. Hence, having made ourselves acquainted with the bands produced by all known metals, if entirely new bands show themselves, it is a proof that an entirely new metal is present in the flame. It is thus that Bunsen and Kirchhoff, the founders of spectrum analysis, discovered Rubidium and Cæsium; and that Thallium, with its superb green band, was discovered by Mr. Crookes.

288. The *permanent gases* when heated to a sufficient temperature, as they may be by the electric discharge, also exhibit characteristic bands in their spectra. By these bands they may be recognized, even at stellar distances.

289. The action of light upon the eye is a test of unrivalled delicacy. In *spectrum analysis* this action is brought specially into play; hence the power of this method of analysis.*

Further Definition of Radiation and Absorption.

290. The terms ray, radiation, and absorption, were employed long prior to the views now entertained regard-

---

* Many persons are incompetent to distinguish one color of the spectrum from another; red and green, for example, are often confounded. Dalton, the celebrated founder of the Atomic Theory, could only distinguish by their form ripe red cherries from the green leaves of the tree. This point is now attended to in the choice of engine-drivers, who have to distinguish one colored signal from another. The defect is called *color-blindness*, and sometimes *Daltonism*.

ing the nature of light. It is necessary more clearly to understand the meaning attached by the undulatory theory to those terms.

291. And to complete our knowledge it is necessary to know that all bodies, whether luminous or non-luminous, are *radiants ;* if they do not radiate light they radiate heat.

292. It is also necessary to know that luminous rays are also heat rays; that the self-same waves of ether falling on a thermometer produce the effects of heat; and impinging upon the retina produce the sensation of light. The rays of greatest heat, however, as already explained, lie entirely without the visible spectrum.

293. The radiation both of light and heat consists in the *communication* of motion from the vibrating atoms of bodies to the ether which surrounds them. The absorption of heat consists in the *acceptance* of motion, on the part of the atoms of a body, from ether which has been already agitated by a source of light or heat. In radiation, then, motion is yielded to the ether; in absorption, motion is received from the ether.

294. When a ray of light or of heat passes through a body without loss; in other words, when the waves are transmitted *through the ether* which surrounds the atoms of the body, without sensibly imparting motion to the atoms themselves, the body is *transparent.* If motion be in any degree transferred from the ether to the atoms, in that degree is the body *opaque.*

295. If either light or radiant heat be absorbed, the absorbing body is *warmed ;* if no absorption takes place, the light or radiant heat, whatever its intensity may be, passes through the body without affecting its temperature.

296. Thus in the dark foci referred to in Note 246, or

in the focus of the most powerful burning mirror which concentrates the beams of the sun, the *air* might be of a freezing temperature, because the absorption of the heat by the air is insensible. A plate of clear rock-salt, moreover, placed at the focus, is scarcely sensibly heated, the absorption being small; while a plate of glass is shivered, and a plate of blackened platinum raised to a white heat, or even fused, because of their powers of absorption.

297. It is here worth remarking that calculations of the temperatures of comets, founded on their distances from the sun, may be, and probably are, entirely fallacious. The comet, even when nearest to the sun, might be intensely cold. It might carry with it round its perihelion the chill of the most distant regions of space. If transparent to the solar rays it would be unaffected by the solar heat, as long as that heat maintained *the radiant form*.

The Pure Spectrum: Fraunhofer's Lines.

298. When a beam of white light issuing from a slit is decomposed, the spectrum really consists of a series of colored images of the slit placed side by side. If the slit be wide, these images *overlap;* but in a *pure* spectrum the colors must not overlap each other.

299. A pure spectrum is obtained by making the slit through which the decomposed beam passes very narrow, and by sending the beam through several prisms in succession, thus augmenting the dispersion.

300. When the light of the sun is thus treated, the solar spectrum is found to be not perfectly continuous; across it are drawn innumerable dark lines, the rays corresponding to which are absent. Dr. Wollaston was the first to observe some of these lines. They were afterward studied with supreme skill by Fraunhofer, who lettered

them and made accurate maps of them, and from him they have been called *Fraunhofer's lines*.

Reciprocity of Radiation and Absorption.

301. To account for the missing rays of the lines of Fraunhofer was long an enigma with philosophers. By the genius of Kirchhoff the enigma was solved. Its solution carried with it a new theory of the constitution of the sun, and a demonstration of a method which enables us to determine the chemical composition of the sun, the stars, and the nebulæ. The application of Kirchhoff's principles by Messrs. Huggins, Miller, Secchi, Janssen, and Lockyer, has been of especial interest and importance.

302. Kirchhoff's explanation of the lines of Fraunhofer is based upon the principle that every body is specially opaque to such rays as it can itself emit when rendered incandescent.

303. Thus the radiation from a carbonic-oxide flame, which contains carbonic acid at a high temparature, is intercepted in an astonishing degree by carbonic acid. If the rays from a sodium flame be sent through a second sodium flame, they will be stopped with particular energy by the second flame. The rays from incandescent thallium vapor are intercepted by thallium vapor, those from lithium vapor by lithium vapor, and so of the other metals.

304. In the language of the undulatory theory, waves of ether are absorbed with special energy—their motion is taken up with special facility—by atoms whose periods of vibration synchronize with the periods of the waves. This is another way of stating that a body absorbs with special energy the rays which it can itself emit.

305. If a beam of white light be sent through the intensely yellow flame of sodium vapor, the yellow con-

stituent of the beam is intercepted by the flame, while rays of other refrangibilities are allowed free transmission.

306. Hence, when the spectrum of the electric light is thrown upon a white screen, the introduction of a sodium flame into the path of the rays cuts off the yellow component of the light, and the spectrum is furrowed by a dark band in place of the yellow.

307. Introducing other flames in the same manner in the path of the beam, if the quantity of metallic vapor in the flame be sufficient, each flame will cut out its own bands. And if the flame through which the light passes contain the vapors of several metals, we shall have the dark characteristic bands of all of them upon the screen.

308. Expanding in idea our electric light until it forms a globe equal to the sun in size, and wrapping round this incandescent globe an atmosphere of flame, that atmosphere would cut off those rays of the globe which it can itself emit, the interception of the rays being declared by dark lines in the spectrum.

309. We thus arrive at a complete explanation of the lines of Fraunhofer, and a new theory of the constitution of the sun. The orb consists of a solid or molten nucleus, in a condition of intense incandescence, but it is surrounded by a gaseous photosphere containing vapors which absorb those rays of the nucleus which they themselves emit. The lines of Fraunhofer are thus produced.

310. The lines of Fraunhofer are narrow bands of *partial* darkness; they are really illuminated by the light of the gaseous envelope of the sun. But this is so feeble in comparison with the light of the nucleus intercepted by the envelope, that the bands appear dark in comparison with the adjacent brilliance.

311. Were the central nucleus abolished, the bands of Fraunhofer *on a perfectly dark ground* would be trans-

formed into a series of bright bands. These would resemble the spectra obtained from a flame charged with metallic vapors. They would constitute the spectrum of the solar atmosphere.

312. It is not necessary that the photosphere should be composed of *pure vapor*. Doubtless it contains vast masses of incandescent cloudy matter, composed of white hot molten particles. These intensely luminous white hot clouds may be the main origin of the light which the earth receives from the sun, and with them the true vapor of the photosphere may be more or less confusedly mingled. But the vapor which produces the lines of Fraunhofer must exist *outside* the clouds, as assumed by Kirchhoff.

Solar Chemistry.

313. From the dark bands of the spectrum we can determine what substances enter into the composition of the solar atmosphere.

314. One example will illustrate the possibility of this. Let the light from the sun and the light from incandescent sodium vapor pass side by side through the same slit, and be decomposed by the same prism. The solar light will produce its spectrum, and the sodium light its yellow band. This yellow band will coincide exactly in position with a characteristic dark band of the solar spectrum, which Fraunhofer distinguishes by the letter D.

315. Were the solar nucleus absent, and did the vaporous photosphere alone emit light, the dark line D would be a bright one. Its character and position prove it to be the light emitted by sodium. This metal, therefore, is contained in the atmosphere of the sun.*

* By reference to note 283 it will be seen that the sodium line is resolved by delicate analysis into a group of lines. The Fraunhofer dark band D is similarly resolved. It ought to be mentioned that both

316. The result is still more convincing when a metal which gives a numerous series of bright bands finds each of its bands exactly coincident with a dark band of the solar spectrum. By this method Kirchhoff, to whom we owe, in all its completeness, this splendid generalization, established the existence of iron, calcium, magnesium, sodium, chromium, and other metals, in the solar atmosphere; and Mr. Huggins has extended the application of the method to the light of the planets, fixed stars, and nebulæ.*

Planetary Chemistry.

317. The light reflected from the moon and planets is solar light; and, if unaffected by the planet's atmosphere, the spectrum of the planet would show the same lines as the solar spectrum.

318. The light of the moon shows no other lines. There is no evidence of an atmosphere round the moon.

319. The lines in the spectrum of Jupiter indicate a powerful absorption by the atmosphere of this planet. The atmosphere of Jupiter contains some of the gases or vapors present in the earth's atmosphere. Feeble lines, some of them identical with those of Jupiter, occur in the spectrum of Saturn.

320. The lines characterizing the atmospheres of Jupiter and Saturn are not present in the spectrum of Mars. The blue portion of the spectrum is mainly the seat of absorption; and this, by giving predominance to the red rays, may be the cause of the red color of Mars.

321. All the stronger lines of the solar spectrum are found in the spectrum of Venus, but no additional lines.

Mr. Talbot and Sir John Herschel clearly foresaw the possibility of employing spectrum analysis in detecting minute traces of bodies.

* Prof. Stokes foresaw the possible application of spectrum analysis to solar chemistry.

Stellar Chemistry.

322. The atmosphere of the star Aldebaran contains hydrogen, sodium, magnesium, calcium, iron, bismuth, tellurium, antimony, mercury. The atmosphere of the star Alpha in Orion contains sodium, magnesium, calcium, iron, and bismuth.

323. No star sufficiently bright to give a spectrum has been observed to be without lines. Star differs from star only in the grouping and arrangement of the numerous fine lines by which their spectra are crossed.

324. The dark absorption lines are strongest in the spectra of yellow and red stars. In white stars the lines, though equally numerous, are very poor and faint.

325. A comparison of the spectra of stars of different colors suggests that the colors of the stars may be due to the action of their atmospheres. Those constituents of the white light of the star on which the lines of absorption fall thickest are subdued, the star being tinted by the residual color.

Father Secchi, of Rome, has studied the light of many hundreds of stars, and has divided them into four classes.

Nebular Chemistry.

326. Some nebulæ give spectra of bright bands, others give continuous spectra. The light from the former emanates from intensely heated matter existing *in a state of gas*. This may in part account for the weakness of the light of these nebulæ.

327. It is probable that two of the constituents of the gaseous nebulæ are hydrogen and nitrogen.

The Red Prominences and Envelope of the Sun.

328. Astronomers had observed during total eclipses of the sun vast red prominences extending from the solar

limb many thousand miles into space. The intense illumination of the circumsolar region of our atmosphere masks, under ordinary circumstances, the red prominences. They are quenched, as it were, by excess of light.

329. But when, by the intervention of the dark body of the moon, this light is cut off, the prominences are distinctly seen.

330. It was proved by Mr. De la Rue and others that the red matter of the prominences was wrapped round a large portion of the sun's surface. According to the observations of Mr. Lockyer, the red matter forms a *complete envelope* round the sun.

331. Examined by the spectroscope the matter of the prominences shows itself to be, for the most part, incandescent hydrogen. With it are mixed the vapors of sodium and magnesium.

332. Mr. Janssen, in India, and Mr. Lockyer subsequently, but independently, in England, proved that the bright bands of the prominences might be seen without the aid of a total eclipse. The explanation of this discovery is glanced at in Note 284, where the intensity of the bright bands of incandescent gases was referred to the practical absence of dispersion.

333. By sending the light, which under ordinary circumstances masks the hydrogen bands, through a sufficient number of prisms, it may be dispersed, and thereby enfeebled in any required degree. When sufficiently enfeebled the undispersed light of the incandescent hydrogen dominates over that of the continuous spectrum. By going completely round the periphery of the sun Mr. Lockyer found this hydrogen atmosphere everywhere present, its depth, generally about 5,000 miles, being indicated by the length of its characteristic bright lines. Where the hydrogen ocean is shallow, the bright bands are short;

where the prominences rise like vast waves above the level of the ocean, the bright lines are long. The prominences sometimes reach a height of 70,000 miles.

The Rainbow.

334. A beam of solar light, falling obliquely on the surface of a rain-drop, is refracted on entering the drop; it is in part reflected at the back of the drop, and on emerging from the drop it is again refracted.

335. By these two refractions on entrance and on emergence the beam of light is decomposed, and it quits the drop resolved into its colored constituents. It is received by the eye of an observer who faces the drop and turns his back to the sun.

336. In general the solar rays, when they quit the drop, are *divergent*, and therefore produce but a feeble effect upon the eye. But at *one* particular angle the rays, after having been twice refracted and once reflected, issue from the drop almost perfectly parallel. They thus preserve their intensity like rays reflected from a parabolic mirror, and produce a corresponding effect upon the eye. The angle at which this parallelism is established varies with the refrangibility of the light.

337. Draw a line from the sun to the observer's eye and prolong this line beyond the observer. Conceive another line drawn from the eye enclosing an angle of 42° 30' with the line drawn to the sun. The rain-drop struck by this second line will send to the eye a parallel beam of *red light*. Every other drop similarly situated, that is to say, every drop at an angular distance of 42° 30' from the line drawn to the sun, will do the same. We thus obtain a *circular band* of red light, forming part of the base of a cone, by which the eye of the observer is the apex. Because

of the angular magnitude of the sun the width of this band will be half a degree.

338. From the eye of the observer conceive another line to be drawn enclosing an angle of 40° 30' with the line drawn to the sun. A drop struck by this line will send along the line an almost perfectly parallel beam of *violet* light to the eye. All drops at the same angular distance will do the same, and we shall obtain a band of violet light of the same width as the red. These two bands constitute the limiting colors of the rainbow, and between them the bands corresponding to the other colors lie.

339. The rainbow is in fact a spectrum, in which the rain-drops play the part of prisms. The width of the bow from red to violet is about two degrees. The size of the arc visible at any time manifestly depends upon the position of the sun. The bow is grandest when it is formed by the rising or the setting sun. An entire semicircle is then seen by an observer on a plain, while from a mountain-top a still greater arc is visible.

340. The angular distances and the order of colors here given correspond to the *primary bow*, but in addition to this we usually see a *secondary bow* of weaker hues, and in which the order of the colors is that of the primary *inverted*. In the primary the red band forms the convex surface of the arch; it is the largest band; in the secondary the violet band is outside, the red forming the concavity of the bow.

341. The secondary bow is produced by rays which have undergone *two* reflections within the drop, as well as two refractions at its surface. It is this double internal reflection that weakens the color. In the primary bow the incident rays strike the upper hemisphere of the drop, and emerge from the lower one; in the secondary bow the incident rays strike the lower hemisphere of the drop, emerge

from the upper one, and then cross the incident rays to reach the eye of the observer. The secondary bow is $3\frac{1}{4}$ degrees wide, and it is $7\frac{1}{4}$ degrees higher than the primary. From the space between the two bows part of the light reflected from the *anterior surfaces* of the rain-drops reaches the eye; but no light whatever that *enters* the rain-drops in this space is reflected to the eye. Hence this region of the falling shower is darkest.

Interference of Light.

342. In wave-motion we must clearly distinguish the motion of the *wave* from the motion of the *individual particles* which at any moment constitute the wave. For while the wave moves forward through great distances, the individual particles of water concerned in its propagation perform a comparatively short excursion to and fro. A sea-fowl, for example, as the waves pass it, is not carried forward, but moves up and down.*

343. Here, as in other cases, the distance through which the individual water particles oscillate, or through which the fowl moves vertically up and down, is called the *amplitude* of the oscillation.

344. When light from two different sources passes through the same ether, the waves from the one source must be more or less affected by the waves from the other. This action is most easily illustrated by reference to water-waves.

345. Let two stones be cast at the same moment into still water. Round each of them will spread a series of circular waves. Let us fix our attention on a point A in the water, equally distant from the two centres of disturbance. The first two crests of both systems of waves reach

* Strictly speaking, the water particles describe *closed curves*, and not straight vertical lines.

this point at the same moment, and it is lifted by their joint action to twice the height that it would attain through the action of either wave taken singly.

346. The first depression, or *sinus* as it is called, of the one system of waves also reaches the point A at the same moment as the first sinus of the other, and through their joint action the point is depressed to twice the depth that it would attain by the action of either sinus taken singly.

347. What is true of the first crest and the first depression is also true of all the succeeding ones. At the point A the successive crests will coincide, and the successive depressions will coincide, the agitation of the point being twice what it would be if acted upon by one only of the systems of waves.

348. The *length of a wave* is the distance from any crest, or any sinus, to the crest or sinus next preceding or succeeding. In the case of the two stones dropped at the same moment into still water, it is manifest that the coincidence of crest with crest and of sinus with sinus would also take place if the distance from the one stone to the point A exceeded the distance of the other stone from the same point by *a whole wave-length*. The only difference would be, that the second wave of the nearest stone would then coincide with the first wave of the most distant one. The one system of waves would here be retarded a whole wave-length behind the other system.

349. A little reflection will also make it clear that coincidence of crest with crest and of sinus with sinus will also occur at the point A when the retardation of the one system behind the other amounts to any number of *whole wave-lengths*.

350. But if we suppose the point A to be *half a wave-length* more distant from the one stone than from the other, then as the waves pass the point A the crests of one

of the systems will always coincide with the sinuses of the other. When a wave of the one system tends to elevate the point A, a wave from the other system will, at the same moment, tend to depress it. As a consequence the point will neither rise nor sink, as it would do if acted upon by either system of waves taken singly. The same neutralization of motion occurs where the difference of path between the two stones and the point A amounts to any *odd* number of half wave-lengths.

351. Here, then, by adding motion to motion, we abolish motion and produce rest. In precisely the same way we can, by adding sound to sound, produce silence, one system of sound-waves being caused to neutralize another. So also by adding heat to heat we can produce cold, while by adding light to light we can produce darkness. It is this perfect identity of the deportment of light and radiant heat with the phenomena of wave-motion that constitutes the strength of the Theory of Undulation.

352. This action of one system of waves upon another, whereby the oscillatory motion is either augmented or diminished, is called *Interference*. In relation to optical phenomena it is called the Interference of Light. We shall henceforth have frequent occasion to apply this principle.

Diffraction, or the Inflection of Light.

353. Newton, who was familiar with the idea of an ether, and indeed introduced it in some of his speculations, objected that if light were propagated by waves, shadows could not exist; for that the waves would bend round opaque bodies, and abolish the shadows behind them. According to the wave theory this bending round of the waves actually occurs, but the different portions

DIFFRACTION, OR THE INFLECTION OF LIGHT. 89

of the inflected waves destroy each other by their interference.

354. This bending of the waves of light round the edges of opaque bodies, receives the name of *Diffraction* or *Inflection* (German, Beugung). We have now to consider some of the effects of diffraction.

355. And for this purpose it is necessary that our source of light should be a physical point or a fine line: for when an extensive luminous surface is employed, the effects of its different points in diffraction phenomena neutralize each other.

356. A *point* of light may be obtained by converging, by a lens of short focus, the parallel rays of the sun, admitted through a small aperture into a dark room. The small image of the sun formed at the focus is here our luminous point. The image of the sun formed on the surface of a silvered bead, or indeed upon the convex furface of a glass lens, or of a watch-glass blackened within, also answers the purpose.

357. A *line* of light is obtained by admitting the sunlight through a slit, and sending the slice of light through a cylindrical lens. The rectangular beam is contracted to a physical line at the focus of the lens. A glass tube blackened within and placed in the light, reflects from its surface a luminous line which also answers the purpose. For many experiments, indeed, the circular aperture, or the slit itself, suffices without any condensation by a .ens.

358. In the experiment now to be described, a slit of variable width is placed in front of the electric lamp, and this slit is looked at from a distance through another slit, also of variable aperture. The light of the lamp is rendered monochromatic by placing a pure red glass in front of the slit.

359. With the eye placed in the straight line drawn through both slits from the incandescent carbon points of the electric lamp an extraordinary appearance is observed. Firstly, the slit in front of the lamp is seen as a vivid rectangle of light; but right and left of it is a long series of rectangles, decreasing in vividness, and separated from each other by intervals of absolute darkness.

360. The breadth of the bands varies with the width of the slit placed in front of the eye. If the slit be widened, the images become narrower, and crowd more closely together; if the slit be narrowed, the images widen and retreat from each other.

361. It may be proved that the width of the bands is inversely proportional to the width of the slit held in front of the eye.

362. Leaving every thing else unchanged, let a blue glass or a solution of ammonia sulphate of copper, which gives a very pure blue, be placed in the path of the light. A series of blue bands is thus obtained, exactly like the former in all respects save one; the blue rectangles are *narrower*, and they are *closer together*, than the red ones.

363. If we employ colors of intermediate refrangibilities between red and blue, which we may do by causing the different colors of a spectrum to shine through the slit, we should obtain bands of color intermediate in width and occupying intermediate positions between those of the red and blue. Hence when *white light* passes through the slit the various colors are not superposed, and instead of a series of monochromatic bands, separated from each other by intervals of darkness, we have a series of colored spectra placed side by side, the most refrangible color of each spectrum being nearest to the slit.

364. When the slit in front of the camera is illuminated by a candle-flame, instead of the more intense electric light,

substantially the same effects, though less brilliant, are observed.

365. What is the meaning of this experiment, and how are the lateral images of the slit produced? Of these and certain accompanying results the emission theory is incompetent to offer any explanation. Let us see how they are accounted for by the theory of undulation.

366. For the sake of simplicity, we will consider the case of monochromatic light. Conceive a wave of ether advancing from the first slit toward the second, and finally filling the second slit. When the wave passes through the latter it not only pursues its direct course to the retina, but diverges right and left, tending to throw into motion the entire mass of the ether behind the slit. In fact, *every point of the wave which fills the slit is itself a centre of new wave-systems, which are transmitted in all directions through the ether behind the slit.* We have now to examine how these secondary waves act upon each other.

367. First, let us regard the central rectangle of the series. It is manifest that the different parts of every transverse section of the wave, which in this case fills our slit, reach the retina at the same moment. They are in complete accordance, for no one portion is retarded in reference to any other portion. The rays thus coming direct from the source through the slit to the retina produce the central band of the series.

368. But now let us consider those waves which diverge *obliquely* from the slit. In this case, the waves from the two edges of the slit have, in order to reach the retina, to pass over *unequal distances*. Let us suppose the difference in path of the two marginal rays to be a whole wave-length of the red light; how must this difference affect the final illumination of the retina?

369. Fix your attention upon the particular ray or line of light that passes exactly through the *centre* of the slit to the retina. The difference of path between this central ray and the two marginal rays is, in the case here supposed, *half a wave-length*. The least reflection will make it clear that every ray on the one side of the central line finds a ray upon the other side, from which its path differs by half an undulation, with which, therefore, it is in complete discordance. The consequence is that the light on the one side of the central line will completely abolish the light on the other side of that line, absolute darkness being the result of their mutual extinction. The first *dark* interval of our series of bands is thus accounted for. It is produced by an obliquity which causes the paths of the marginal rays to be *a whole wave-length* different from each other.

370. When the difference between the paths of the marginal rays is *half a wave-length*, a *partial* destruction of the light is effected. The luminous intensity corresponding to this obliquity is a little less than one-half—accurately 0.4—of that of the undiffracted light.

371. If the paths of the marginal rays be three semi-undulations different from each other, and if the whole beam be divided into three equal parts, two of these parts will completely neutralize each other, the third only being effective. Corresponding, therefore, to an obliquity which produces a difference of three semi-undulations in the marginal rays, we have a luminous band, but one of considerably less intensity than the undiffracted central band.

372. With a marginal difference of path of four semi-undulations we have a second extinction of the entire beam, a space of absolute darkness corresponding to this obliquity. In this way we might proceed further, the

general result being that, whenever the obliquity is such as to produce a marginal difference of path of an *even* number of semi-undulations, we have complete extinction; while, when the marginal difference is an *odd* number of semi-undulations, we have only partial extinction, a portion of the beam remaining as a luminous band.

373. A moment's reflection will make it plain that the shorter the wave, the less will be the obliquity required to produce the necessary retardation. The maxima and minima of blue light must, therefore, fall nearer to the centre than the maxima and minima of red light. The maxima and minima of the other colors fall between these extremes. In this simple way the undulatory theory completely accounts for the extraordinary appearance referred to in Note 359. When a slit and telescope are used, instead of the slit and naked eye, the effects are magnified and rendered more brilliant.

Measurement of the Waves of Light.

374. We are now in a condition to solve the important problem of measuring *the length* of a wave of light.

375. The first of our dark bands corresponds, as already explained, to a difference of marginal path of one undulation; our second dark band to a difference of path of two undulations; our third dark band to a difference of three undulations, and so forth. With a slit 1.35* millimetre wide, Schwerd found the angular distance of the first dark band from the centre of the field to be 1' 38''. The angular distances of the other dark bands are twice, three times, four times, etc., this quantity, that is to say, they are *in arithmetical progression.*

376. Draw a diagram of the slit E C with the beam

* The millimetre is about $\frac{1}{25}$ of an inch.

passing through it at the obliquity corresponding to the first dark band. Let fall a perpendicular from one edge, E, of the slit on the marginal ray of the other edge at d. The distance, $c\,d$, between the foot of this perpendicular and the other edge is the length of the wave of light. From the centre E, with the width E C as radius, suppose a semicircle to be described; its radius being 1.35, the length of this semicircle is readily found to be 4.248 millimetres. Now, the length of this semicircle is to the length $c\,d$ of the wave as 180° to 1′ 38″, or as 648,000″ to 98″. Thus we have the proportion—

648,000 : 98 :: 4.248 to the wave-length $c\,d$.*

Making the calculation, we find the wave-length for this particular kind of light (red), to be 0.000643 of a millimetre, or 0.000026 of an inch.

377. Instead of receiving them directly upon the retina, the colored fringes may be received upon a screen. In this case it is desirable to employ a lens of considerable convergent power to bring the beam from the first slit to a focus, and to place the second slit or other diffracting edge or edges between the focus and the screen. The light in this case virtually emanates from the focus.

378. If the edge of a knife be placed in the beam parallel to the slit, the shadow of the edge upon the screen will be bounded by a series of parallel colored fringes. If the light be monochromatic the bands will be simply bright and dark. The back of the knife produces the same effect as its edge. A wooden or an ivory paper-knife produces precisely the same effect as a steel knife. The fringes are absolutely independent of the character of the substance round the edge of which the light is diffracted.

* $C\,d$ is so minute that it practically coincides with the circle drawn round E.

379. A thick wire placed in the beam has colored fringes on each side of its shadow. If the wire be *fine*, or if a human hair be employed, the geometric shadow itself will be found occupied by parallel stripes. The former are called the *exterior fringes*, the latter the *interior fringes*. In the hands of Young and Fresnel all these phenomena received their explanation as effects of interference.

380. A *slit* consists of two edges facing each other. When a slit is placed in the beam between the focus and the screen, the space between the edges is occupied by stripes of color.

381. Looking at a distant point of light through a small circular aperture the point is seen encircled by a series of colored bands. If monochromatic light be used these bands are simply bright and dark, but with white light the circles display iris-colors.

382. These results are capable of endless variation by varying the size, shape, and number of the apertures through which the point of light is observed. The street lamps at night, looked at through the meshes of a handkerchief, show diffraction phenomena. The diffraction effects obtained by Schwerd in looking through a bird's feathers are very gorgeous. The iridescence of Alpine clouds is also an effect of diffraction.*

---

* This may be imitated by the spores of Lycopodium. The diffraction phenomena of "actinic clouds" are exceedingly splendid. One of the most interesting cases of diffraction by small particles that ever came before me was that of an artist whose vision was disturbed by vividly-colored circles. When he came to me he was in great dread of losing his sight; assigning as a cause of his increased fear that the circles were becoming larger and the colors more vivid. I ascribed the colors to minute particles in the humors of the eye, and encouraged him by the assurance that the increase of size and vividness indicated that the diffracting particles were becoming *smaller*, and that they might finally be altogether absorbed. The prediction was verified. It is needless to say

383. Following out the indications of theory, Poisson was led to the paradoxical result that in the case of *an opaque circular disk* the illumination of the centre of the shadow, caused by diffraction at the edge of the disk, is precisely the same as if the disk were altogether absent. This startling consequence of theory was afterward verified experimentally by Arago.

Colors of Thin Plates.

284. When a beam of monochromatic light—say of pure red, which is most easily obtained by absorption—falls upon a thin, transparent film, a portion of the light is reflected at the first surface of the film; a portion enters the film, and is in part reflected at the second surface.

385. This second portion having crossed the film to and fro is *retarded* with reference to the light first reflected. The case resembles that of our two stones dropped into still water at unequal distances from the point A (Note 345).

386. If the thickness of the film be such as to retard the beam reflected from the second surface a whole wavelength, or any number of whole wave-lengths—or, in other words, any *even* number of half wave-lengths—the two reflected beams, travelling through the same ether, will be in *complete accordance;* they will therefore support each other, and make the film appear brighter than either of them would do taken singly.

387. But if the thickness of the film be such as to retard the beam reflected from the second surface half a wavelength, or any *odd* number of half-wave lengths, the two reflected beams will be in *complete discordance;* and a destruction of light will follow. By the addition of light

one word on the necessity of optical knowledge in the case of the practical oculist.

which has undergone more than one reflection at the second surface to the light which has undergone only one reflection, the beam reflected from the first surface may be *totally* destroyed. Where this total destruction of light occurs the film appears black.

388. If the film be of variable thickness, its various parts will appear bright or dark, according as the thickness favors the accordance or discordance of the reflected rays.

389. Because of the different lengths of the waves of light, the different colors of the spectrum require different thicknesses to produce accordance and discordance; the longer the waves, the greater must be the thickness of the film. Hence those thicknesses which effect the extinction of one color will not effect the extinction of another. When, therefore, a film of variable thickness is illuminated by *white* light, it displays a variety of colors.

390. These colors are called the colors of *thin plates*.

391. The colors of the soap-bubble; of oil or tar upon water; of tempered steel; the brilliant colors of lead skimmings; Nobili's metallo-chrome; the flashing colors of certain insects' wings, are all colors of thin plates. The colors are produced by transparent films of all kinds. In the bodies of crystals we often see iridescent colors due to vacuous films produced by internal fracture. In cutting the dark ice under the moraines of glaciers internal fracture often occurs, and the colors of thin plates flash forth from the body of the ice with extraordinary brilliancy.

392. Newton placed a lens of small curvature in optical contact with a plane surface of glass. Between the lens and the surface he had a film of air, which gradually augmented in thickness from the point of contact outward. He thus obtained in monochromatic light a series of bright and dark *rings*, corresponding to the different thick-

nesses of the film of air, which produced alternate accordance and discordance.

393. The rings produced by violet he found to be smaller than those produced by red, while the rings produced by the other colors fell between these extremes. Hence when white light is employed, "Newton's Rings" appear as a succession of circular bands of color. A far greater number of the rings is visible in monochromatic than in white light, because the differently-colored rings, after a certain thickness of film has been attained, become superposed and reblended to form white light.

394. Newton, considering the means at his disposal, measured the diameters of his rings with marvellous accuracy; he also determined from its focal length and its refractive index the diameter of the sphere of which his lens formed a part. He found the squares of the diameters of his rings to be in *arithmetical progression*, and consequently that the *thicknesses* of the film of air corresponding to the diameters of the rings were also in arithmetical progression.

395. He determined the *absolute thicknesses* of the plates of air at which the rings were formed. Employing the most luminous rays of the spectrum, that is, the rays at the common boundary of the yellow and orange, he found the thickness corresponding to the first bright ring to be $\frac{1}{178000}$ of an inch.

396. The entire series of bright rings were formed at the following successive thicknesses:

$$\frac{1}{178000}, \frac{3}{178000}, \frac{5}{178000}, \frac{7}{178000}, \text{etc.},$$

and the series of dark rings, separating the bright ones, at the thicknesses

$$\frac{2}{178000}, \frac{4}{178000}, \frac{6}{178000}, \frac{8}{178000}, \text{etc.}$$

397. To account for the rings, Newton assumed that the light particles were endowed with *fits of easy transmis-*

sion and of easy reflection. He probably figured those particles as endowed at the same time with a motion of translation through space, and a motion of rotation round their own axes. If we suppose such particles to resemble little magnets which present alternately attractive and repulsive poles to the surface which they approach, we have a conception in conformity with the notion of Newton.

398. According to this conception ordinary reflection and refraction would depend upon the presentation of the repulsive or the attractive poles of the particles to the reflecting or refracting surface.

399. Figure then the rotating light particles entering the film of air between Newton's lens and plate. If the distance between both be such as to enable the light particle to perform a *complete rotation*, it will present *at the second surface* of the film of air the same pole that it presented at the first. It will therefore be *transmitted*, and will not return to the eye.

400. This effect would also take place if the distance between the plate and lens were such as to enable the light particle to perform two, three, four, etc., complete rotations. *The dark rings of Newton were thus accounted for.* They occurred at places where the light particles, instead of being sent back to the eye from the second surface of the film, were transmitted through that surface.

401. But if the thickness of the film be such as to allow the light particle which has entered the first surface to perform only *half a rotation* before it arrives at the second surface; then a repulsive pole will be presented to the latter, and the particle will be driven back to the eye. The same will occur if the distance be such as to enable the light particle to perform three, or five, or seven, etc., semi-rotations. *The bright rings of Newton were thus*

accounted for; they occurred at places where the light particles on reaching the second surface of the film were reflected back to the eye.

402. The theory of emission is here at direct issue with the theory of undulation. Newton assumes that the action which produces the alternate bright and dark rings takes place at a *single* surface; i.e., the second surface of the film. The undulatory theory affirms that the rings are caused by the interference of rays reflected from *both* surfaces. This has been proved to be the case. By employing polarized light (to be subsequently described and explained) we can destroy the reflection at the first surface of the film, and when this is done the rings vanish altogether.

403. The beauty and subtlety of Newton's conception are, however, manifest; and the theory was apparently supported by the fact that rings of feeble intensity are actually formed by *transmitted light*, and that the bright rings by transmitted light correspond to thicknesses which produce dark rings in reflected light.

404. The transmitted rings are referred by the undulatory theory to the interference of rays which have passed directly through the film, with others which have undergone *two reflections* within the film. They are thus completely accounted for.

Note.—The thickness $\frac{1}{178000}$ of an inch referred to in Note 396, as that corresponding to the first bright ring, is *one-fourth* of the length of an undulation of the light employed by Newton. Hence, in passing to and fro through the film, the rays reflected at the second surface are *half* an undulation behind those reflected at the first surface. At this thickness, therefore, the ring ought, according to the principles of interference, to be *dark* instead of *bright.* The same remarks apply to the thicknesses $\frac{3}{178000}$,

DOUBLE REFRACTION. 101

$\frac{5}{118000}$, etc.; the former corresponds to a retardation of three, and the latter to a retardation of five semi-undulations. With regard to the *dark* rings, the first of them occurs at a thickness the double of which is the length of a whole undulation; the second of them occurs at a thickness which, when doubled, is equal to two wave-lengths; the third at a thickness the double of which is three wave-lengths. Hence, if we take *the thickness of the film alone* into account, the bright rings ought to be dark, and the dark rings bright.

But something besides thickness is to be considered here. In the case of the first surface of the film the wave passes from the dense ether of the glass into the rare ether of the air. In the case of the second surface of the film the wave passes from the rare ether of the air into the dense ether of the glass. This difference at the two reflecting surfaces of the film can be proved to be equivalent *to the addition of half a wave-length* to the thickness of the film. To the absolute thickness, therefore, as measured by Newton, half a wave-length is in each case to be added; when this is done the rings follow each other in exact accordance with the law of interference enunciated in Notes 348 to 350.

Double Refraction.

405. In air, water, and well-annealed glass, the luminiferous ether has the same elasticity in all directions. There is nothing in the molecular grouping of these substances to interfere with the perfect homogeneity of the ether.

406. But when water crystallizes to ice, the case is different; here the molecules are constrained by their proper forces to arrange themselves in a certain determinate manner. They are, for example, closer together in

some directions than in others. This arrangement of the molecules carries along with it an arrangement of the surrounding ether, which causes it to possess *different degrees of elasticity in different directions*.

407. In a plate of ice, for example, the elasticity of the ether in a direction perpendicular to the surface of freezing is different from its elasticity in a direction parallel to the same surface.

408. This difference is displayed in a peculiarly striking manner by Iceland spar, which is crystallized carbonate of lime; and in consequence of the existence of these two different elasticities, a wave of light passing through the spar *is divided into two;* the one rapid, corresponding to the greater elasticity, and the other slow, corresponding to the lesser elasticity.

409. Where the velocity is greatest, the refraction is least; and where the velocity is least the refraction is greatest. Hence in Iceland spar, as we have two waves moving with different velocities, we have *double refraction*.

410. This is also true of the greater number of crystalline bodies. If the grouping of the molecules be not in all directions alike, the ether will not be in all directions equally elastic, and double refraction will infallibly result.

411. In rock-salt, alum, and other crystals, this homogeneous grouping of the molecules actually occurs, and such crystals behave like glass, water, or air.

412. In certain doubly refracting crystals the molecules are arranged in the same manner on all sides of a certain direction. For example, in the case of ice the molecular arrangement is the same all round the perpendiculars to the surface of freezing.

413. In like manner, in Iceland spar the molecules are

arranged symmetrically round the crystallographic axis, that is, round the shortest diagonal of the rhomb into which the crystal may be cloven.*

414. When a beam of light passes through ice perpendicular to the surface of freezing, or through Iceland spar parallel to the crystallographic axis, *there is no double refraction.* These cases are representative; that is to say, there is no double refraction in the direction round which the molecular arrangement is in all directions the same.

415. This direction of no double refraction is called *the optic axis* of the crystal.

NOTE.—The vibrations of the ether being *transverse* to the direction of the ray, the elasticity which determines the rapidity of transmission is that at *right angles* to the ray's direction. In Iceland spar the velocity is slowest in the direction of the axis; hence the elasticity at right angles to the axis is a *minimum.* The ray, on the other hand, whose vibrations are executed along the axis is the most rapid; hence the elasticity of the ether along the axis is a *maximum.* In perfectly homogeneous bodies the surface of elasticity would be spherical; it would be measured by the same length of radius in all directions. In the case of Iceland spar the surface of elasticity is an ellipsoid whose longer axis coincides with the axis of the crystal.

* The arrangement of the molecules is such, that Iceland spar may be cloven with great and equal facility in three different directions. The *planes of cleavage* are here oblique to each other. Rock-salt also cleaves readily and equally in three directions, the planes of cleavage being at right angles to each other. Hence, while rock-salt cleaves into *cubes*, Iceland spar cleaves into *rhombs.* Many crystals cleave with different facilities in different directions. Selenite and crystallized sugar (sugar-candy) are examples.

Phenomena presented by Iceland Spar.

416. The two beams into which the incident beam is divided by the spar do not behave alike. One of them obeys the ordinary law of refraction; its index of refraction is perfectly constant and independent of its direction through the crystal. The angles of incidence and refraction are in the same plane, as in the case of ordinary refraction. The ray which behaves thus is called *the ordinary ray*. In its case the sine of the angle of incidence is to the sine of the angle of refraction, or the velocity of light in air is to its velocity in the crystal, in the constant ratio of 1.654 to 1. The number 1.654 is *the ordinary index* of Iceland spar.

417. But the other beam acts differently. Its index of refraction is not constant, nor is the angle of refraction as a general rule in the same plane as the angle of incidence. The ray which behaves thus is called *the extraordinary ray*. If a prism be formed of the spar with its refracting angle parallel to the optic axis, when the incident beam traverses the prism *at right angles to the optic axis*, the separation of its two parts is a *maximum*. Here the full difference of elasticity between the axial direction and that perpendicular to it comes into play, and the extraordinary ray suffers its *minimum* retardation, and therefore its minimum refraction. Its refractive index is then 1.483.

418. The index of refraction of the extraordinary ray varies with its direction through the crystal from 1.483 to 1.654. The *minimum value* of the ratio of the two sines, or of the two velocities, viz., 1.483, is called *the extraordinary index*.

419. When a small aperture through which light passes is regarded through a rhomb of Iceland spar two apertures are seen. If the rhomb be placed over a black dot on a sheet of white paper, two dots will be seen; and if the spar be turned, one of the images of the aperture or of the dot will rotate round the other.

420. The rotating image is that formed by the extraordinary ray.

421. One of the two images of the dot is also *nearer* than the other. The ordinary ray behaves as if it came from a more highly refractive medium, and the greater the refraction the nearer must the image appear. The apparent shallowness of water is referred to in Notes 131 and 132. With bisulphide of carbon the shallowness would be more pronounced, because the refraction is greater. In Iceland spar the ordinary index bears nearly the same relation to the extraordinary as the index of bisulphide of carbon to that of water; hence the *ordinary image* must appear nearer than the extraordinary one.

422. Brewster showed that a great number of crystals possessed *two* optic axes, or two directions on which a beam passes through the crystal without division. Crystallized sugar, mica, heavy spar, sulphate of lime, and topaz, are examples.

423. Thus crystals divide themselves into—

I. *Single refracting crystals*, such as rock-salt, alum, and fluor-spar; and

II. *Double refracting crystals*, of which we have two kinds, viz.:

a. Uniaxal crystals, or those with a single optic axis, such as Iceland spar, rock-crystal, and tourmaline; and—

b. Biaxal crystals, or those which possess two optic axes, such as arragonite, felspar, and those mentioned in 422.

424. When on a plate of Iceland spar cut perpendicular to the axis, a beam of light falls obliquely, the ordinary ray being the more refracted is nearer to the axis than the extraordinary. The extraordinary ray is as it were *repelled* by the axis. But Biot showed that there are many crystals in which the reverse occurs, in which, that is to say, the extraordinary ray is nearer to the axis than the ordinary, being as it were *attracted*. The former class he called repulsive or *negative* crystals; Iceland spar, ruby, sapphire, emerald, beryl, and tourmaline, being examples. The latter class he called attractive or *positive* crystals, rock-crystal, ice, zircon, being examples.

The Polarization of Light.

425. The double refraction of Iceland spar was discovered by Erasmus Bartholinus, and was first described by him in a work published in Copenhagen in 1669. The celebrated Huyghens sought to account for the phenomenon on the principles of a wave theory, and he succeeded in doing so.

426. In his experiments on this subject, Huyghens found that when a common luminous beam passes through Iceland spar in any direction save one (that of the optic axis), it is always divided into two beams of *equal intensity*; but that when *either* of these two half-beams is sent through a second piece of spar, it is usually divided into two of *unequal intensity*; and that there are two positions of the spar in which one of the beams vanishes altogether.

427. On turning the spar round this position of absolute disappearance, the missing beam appeared; its companion at the same time becoming dimmer; both of them then passed through a phase of equal intensity, and when the

rotation was continued, the beam which was first transmitted disappeared.

428. Reflecting on this experiment Newton came to the conclusion that the divided beam had acquired *sides* by its passage through the Iceland spar, and that its interception and transmission depended on the way on which those sides presented themselves to the molecules of the second crystal. He compared this *two-sidedness* of a beam of light to the *two-endedness* of a magnet known as its polarity; and a luminous beam exhibiting this two-sidedness was afterward said to be *polarized*.

429. In 1808, Malus, while looking through a birefracting prism at one of the windows of the Luxembourg Palace, from which the solar light was reflected, found that in a certain position of the spar, the ordinary image of the window almost wholly disappeared; while, in a position perpendicular to this, the extraordinary image disappeared. He discerned the analogy between this action and that discovered by Huyghens in Iceland spar, and came to the conclusion that the effect was due to some new property impressed upon the light by its reflection from the glass.

430. What is this property? It may be most simply studied and understood by means of the crystal called tourmaline. This crystal is birefractive; it divides a beam of light incident upon it into two, but its molecular grouping, and the consequent disposition of the ether within it, are such that one of these beams is rapidly quenched, while the other is transmitted with comparative freedom.

431. It is to be borne in mind that the motions of the individual ether particles are transverse to the direction in which the light is propagated (read Note 219). *In a*

beam *of ordinary light the vibrations occur in all directions round the line of propagation.*

432. The change suffered by light in passing through a plate of tourmaline, of sufficient thickness, and cut parallel to the axis is this: All vibrations save those executed *parallel to the axis* are quenched within the crystal. Hence the beam emergent from the plate of tourmaline has all its vibrations reduced to a single plane. In this condition it is a beam *of plane polarized light.*

433. Imagine a cylindrical beam of light with all its ether particles vibrating in the same direction—say *horizontally*—looked down upon vertically, the ether particles, if large enough, would be seen performing their excursions to and fro across the direction of the beam. Looked at crosswise horizontally, the particles would be seen advancing and retreating, but their paths would be invisible, every ether particle covering its own path. In the one case we should see *the lines* of excursion; in the other case, *the ends* of the lines only. In this, according to the undulatory theory, consists the *two-sidedness* discovered by Huyghens, and commented on by Newton.

Polarization of Light by Reflection.

434. The quality of two-sidedness is also impressed upon light by reflection. This is the great discovery of Malus. A beam reflected from glass is in part polarized *at all oblique incidences*, a portion of its vibrations being reduced to a common plane. At one particular incidence the beam is *perfectly polarized*, *all* its vibrations being reduced to the same plane. The angle of incidence which corresponds to this perfect polarization is called the *polarizing angle.*

435. The polarizing angle is connected with the index

ot refraction of the medium by a very beautiful law discovered by Sir David Brewster.* When a luminous beam is incident upon a transparent substance, it is in part reflected and in part refracted. At one particular incidence the reflected and refracted portions of the beam are *at right angles to each other*. The angle of incidence is *then* the polarizing angle. This is the geometrical expression of the law of Brewster.

436. The polarizing angle augments with the refractive index of the medium. For water it is 53°, for glass 58°, and for diamond 68°.

437. Thus a beam of ordinary light, whose vibrations are executed in all directions, impinging upon a plate of glass at the polarizing angle, has, after reflection, all its vibrations reduced to a common plane. The direction of the vibrations of the polarized beam *is parallel to the polarizing surface.*

438. Let a beam thus polarized by reflection at the surface of one plate of glass impinge upon a second plate *at the polarizing angle.* In one position of this plate the beam suffers its maximum reflection. In a certain other position the beam is *wholly transmitted,* there is no reflection. In this experiment the angle of incidence remains unchanged, nothing being altered save *the side* of the ray which strikes the reflecting surface.

439. The reflection of the polarized beam is a maximum when the lines along which the ether particles vibrate are *parallel* to the reflecting surface. It is wholly transmitted when the lines of vibration strike the reflecting surface at the polarizing angle. The reflection is then zero. By taking advantage of this fact, the reflection from the first surface of a thin film has been abolished,

* The index of refraction of the medium is the tangent of the polarizing angle.

Newton's rings being thereby rendered incapable of formation, as stated in Note 402.

440. A beam which meets the first surface of a plate of glass with parallel sides at the polarizing angle meets the second surface also at its polarizing angle, and is in part reflected there perfectly polarized. Hence, by augmenting the number of plates, the repeated reflections at their limiting surfaces furnish a polarized beam of greater intensity than that obtained by reflection at a single surface.

Polarization of Light by Refraction.

441. We have hitherto directed our attention to the *reflected* portion of the beam; but the *refracted* portion, which enters the glass, is also partially polarized. The quantities of polarized light in the reflected and refracted beams *are always equal to each other.*

442. The plane of vibration in the refracted beam *is at right angles* to that in the reflected beam.

443. When several plates of glass are placed parallel to each other, and a beam is permitted to fall upon them at the polarizing angle, at every passage from plate to plate a portion of the light is reflected polarized, an equal portion of polarized light entering the glass at the same time. By duly augmenting the number of plates, the polarization by the successive refractions may be rendered sensibly *perfect.* When this occurs, if any further plates be added to the bundle, reflection *entirely ceases* at their limiting surfaces, the beam afterward being wholly transmitted.

Polarization of Light by Double Refraction.

444. In the case last considered the light was polarized by ordinary refraction. The polarization of light by double

refraction has been already touched upon in Notes 432 and 433. We shall now extend our examination of the crystal of tourmaline there referred to, and turn it to account in the examination of other crystals.

445. If a beam of light which has passed through one plate of tourmaline impinge upon a second plate, it will pass through both, if the axes of the two plates be *parallel.* But if they are *perpendicular* to each other, then the light transmitted by the one is quenched by the other, darkness marking the space where the two plates are superposed.

446. If the two axes be *oblique* to each other, a portion of the light will pass through both plates. For, in a manner similar to the resolution of forces in ordinary mechanics, an oblique vibration may be resolved into two, one parallel to the axis of the tourmaline, the other perpendicular to the axis. The latter component is *quenched*, but the former is *transmitted.*

447. Hence if the axes of two plates of tourmaline be perpendicular to each other, a third plate of tourmaline introduced *obliquely* between them, or a plate of any other crystal which acts in a manner similar to the tourmaline, will transmit a portion of the light emergent from the first crystal. The plane of vibration of this light being oblique to the axis of the second crystal, a portion of the light will also pass through the latter. By the introduction, therefore, of a third crystal, with its axis oblique, we abolish in part the darkness of the space where the two rectangular plates are superposed.

Examination of Light transmitted through Iceland Spar.

448. We have now to examine, by means of a plate of tourmaline, the two parts into which a luminous beam is divided in its passage through Iceland spar.

449. Confining our attention to one of the two beams, it is immediately found that in a certain position of the plate the light is freely transmitted, while in the perpendicular position it is completely stopped. This proves the beam emergent from the spar to be *polarized*.

450. From the position of the tourmaline we can immediately infer the direction of vibration in the polarized beam. If transmission occur when the axis of the plate of tourmaline is vertical, the vibrations are vertical; if transmission occur when the tourmaline is horizontal, the vibrations are horizontal. The same mode of investigation teaches us that the second beam emergent from the spar is also polarized.

451. The vibrations of the ether particles in the two beams are executed in planes which are *at right angles to each other*. If the vibrations in the one beam be vertical, in the other they are horizontal. A plate of tourmaline with its axis vertical transmits the former and quenches the latter; while the same plate held horizontally, quenches the former and transmits the latter.

452. A tourmaline plate placed with its axis vertical, in front of the electric lamp, has its image cast by a lens upon a screen. A piece of Iceland spar, with one of its planes of vibration horizontal and the other vertical, placed in front of the lens divides the beam into two, and yields *two images* of the tourmaline. One of these images is *bright*, the other is *dark*. The reason is, that in the light emergent from the tourmaline the vibrations are vertical, and they can only be transmitted through the spar in company with *its* vertically vibrating beam. In the horizontally vibrating beam the tourmaline must appear black.

453. It is also black if the light emergent from it, and surrounding it, meet, at the polarizing angle, a plate of

glass whose plane of reflection is *vertical;* while it is bright when the light is reflected *horizontally.* These effects are consequences of the law of polarization by reflection.

454. Not only do crystallized bodies possess this power of double refraction and polarization; but all bodies whose atomic grouping is such as to cause the ether within them to possess different elasticities in different directions do the same.

455. Thus organic structures are usually double refracting. A double refracting structure may also be conferred on ordinary glass by either strain or pressure. Strains and pressures due to unequal heating also produce double refraction. Unannealed glass behaves like a crystal. A plate of common window-glass, which under ordinary circumstances shows no trace of double refraction, if heated at a single point, is rendered doubly refractive by the strains and pressures propagated round the heated point. The introduction of any of these bodies between the *crossed plates* of tourmaline partly abolishes the darkness caused by the superposition of the plates.

456. Two plates of tourmaline, between which bodies may be introduced and examined by polarized light, constitute a simple form of the *polariscope.* The plate at which the light first enters is called the polarizer, while the second plate is called the analyzer.

457. But the tourmalines are small, usually colored, and under no circumstances competent to furnish an intense beam of polarized light. If one of the parts into which a prism of Iceland spar divides a beam of light could be abolished, the remaining beam would be polarized, and, because of the transparency of the spar, it would be far more intense than any beam obtainable from tourmaline.

458. This has been accomplished with great skill by Nicol. He cut a long parallelopiped of spar into two by a very oblique section; polished the two surfaces, and united them by Canada balsam. The refrangibility of the balsam lies between those of the ordinary and the extraordinary rays in Iceland spar, being less than the former and greater than the latter. When, therefore, a beam of light is sent along the parallelopiped, the ordinary ray, to enter the balsam, must pass from *a denser to a rarer medium*. In consequence of the obliquity of its incidence *it is totally reflected*, and is thus got rid of. The extraordinary ray, on the contrary, in passing from the spar to the balsam passes from a rarer to a denser medium, and is therefore *transmitted*. In this way we obtain a single intense beam of polarized light (read Notes 123, 141, and 142).

459. A parallelopiped prepared in the fashion here described is called a *Nicol's prism*.

460. Nicol's prisms are of immense use in experiments on polarization. With them the best polariscopes are constructed. Reflecting polariscopes are also constructed, consisting of two plates of glass, one of which polarizes the light by reflection, the other examining the light so polarized. The beam reflected from the polarizer is in this case reflected or quenched by the analyzer according as the planes of reflection of the two mirrors are parallel or at right angles to each other.

Colors of Double-refracting Crystals in Polarized Light.

461. A large class of these colors may be illustrated and explained by reference to the deportment of thin plates of gypsum (crystallized sulphate of lime, commonly called selenite) between the polarizer and analyzer of the polariscope.

COLORS OF DOUBLE-REFRACTING CRYSTALS. 115

462. The crystal cleaves with great freedom in one direction; it cleaves with less freedom in two others; the latter two cleavages are also unequal. In other words, gypsum possesses three planes of cleavage, no two of which are equal in value, but one of which particularly signalizes itself by its perfection.

463. By following these three cleavages it is easy to obtain from the crystal diamond-shaped laminæ of any required thinness.

464. The crystal, as might be expected from the character of its cleavages, is double-refracting. A beam of ordinary light impinging at right angles on a plate of gypsum, whose surfaces are those of most perfect cleavage, has its vibrations reduced to two planes at right angles to each other; that is to say, the beam whose ether, prior to entering the gypsum, vibrates in all transverse directions, after it has entered the gypsum, and after its emergence from it, vibrates in two rectangular directions only.

465. The elasticity of the ether is different in these two rectangular directions; consequently the one beam passes more rapidly through the gypsum than the other.

466. In refracting bodies generally the retardation of the light consists in a *diminution of the wave-length* of the light. *The rate of vibration* is unchanged during the passage of the light through the refracting body. The case is exactly similar to that of a musical sound transmitted from water into air. The velocity is reduced to one-fourth by the transfer, because the wave-length is reduced to one-fourth. But the *pitch*, depending as it does on the number of waves which reach the ear in a second, is unaltered.

467. Because of the difference of elasticity between the two rectangular directions of vibration in gypsum, the

waves of ether in the one direction *arc more shortened* than in the other.

468. In the experiments with a plate of gypsum now to be described and explained, we shall employ as polarizer a piece of Iceland spar, one of whose beams is intercepted by a diaphragm. A Nicol's prism shall be our analyzer.

469. When the planes of vibration of the spar and of the Nicol coincide, the light passes through both and may be received upon a screen. When the planes of vibration are at right angles to each other, the light emergent from the spar is intercepted by the Nicol, and the screen is dark.

470. If a plate of selenite be placed between the polarizer and analyzer, with either of its planes of vibration *coincident* with that of the polarizer or analyzer, it produces no change upon the screen. If the screen be light, it remains light; if it be dark, it remains dark after the introduction of the gypsum, which here behaves like a plate of ordinary glass.

471. Let us assume the screen to be dark. Interposing a *thick* plate of gypsum with its directions of vibration *oblique* to that of the polarizer or analyzer, *white* light reaches the screen. If the plate be *thin*, the light which reaches the screen is *colored*. If the plate be of uniform thickness, the color is uniform. If of different thicknesses, or if in cleaving thin scales cling to the surface of the film, some portions of the plate will be differently colored from the rest.

472. When thick plates are employed, the different colors, as in the case of thin plates, are superposed, and reblended to white light.

473. The quantity of light which reaches the eye is a maximum when the planes of vibration of the gypsum

COLORS OF DOUBLE-REFRACTING CRYSTALS. 117

enclose an angle of 45° with those of the polarizer and analyzer.

474. If the plate of selenite be a thin wedge, and if the light be monochromatic, say red, alternately bright (red) and dark bands are thrown upon the screen.

475. If, instead of red light, *blue* be employed, the blue bands are found to occur at smaller thicknesses than those which produced the red: other colors occur at intermediate thicknesses. Hence when *white* light is employed, instead of bands of brightness separated from each other by bands of darkness, we have a series of iris-colored bands.

476. If, instead of a wedge gradually augmenting in thickness from the edge toward the back, we employ a disk gradually augmenting in thickness from the centre outward; instead of a series of parallel bands we obtain under similar circumstances, in *white* light, a series of concentric iris-colored circles.

477. Here, then, we have in the first instance a beam of plane polarized light impinging on the selenite. The direction of vibration of this beam is resolved into two others at right angles to each other; namely, into the two directions in which the ether vibrates within the crystal. One of these systems of waves is *retarded* with reference to the other.

478. But as long as the rays vibrate *at right angles to each other*, they cannot interfere so as to augment or diminish the intensity. To effect such interference the rays must vibrate *in the same place*.

479. The function of the analyzer is to reduce the two rectangular wave-systems to a single plane. Here the effect of retardation is at once felt, and the waves conspire or oppose each other according as their vibrations are *in the same phase* or in *opposite phases*.

480. When the vibration planes of the polarizer and analyzer are *parallel*, a thickness of the gypsum crystal which produces a retardation of *half an undulation* causes the light to be extinguished by the analyzer.

481. When the polarizer and analyzer are *crossed*, a retardation of half an undulation, or of any odd number of half undulations, within the crystal does not produce extinction when these vibrations are compounded by the analyzer. A retardation of a whole undulation, or of any number of whole undulations, produces in this case extinction. This, when followed out, is a plain consequence of the composition of the vibrations.

482. Expressed generally, the phenomena exhibited by the parallel and crossed polarizer and analyzer are *complementary*. If the field be dark when they are crossed, it is bright when they are parallel. If the field be green when they are crossed, it is red when they are parallel; if yellow when they are crossed, it is blue when they are parallel. Thus a rotation of 90° always brings out the complementary color.

483. If instead of the Nicol we employ a birefracting prism of Iceland spar, the colors of the selenite produced by the two oppositely-polarized beams will be complementary. The overlapping of the two colors always produces *white*. Any other double-refracting substance, whether crystallized, organized, mechanically pressed or strained, exhibits, on examination by polarized light, phenomena similar to those of the gypsum.

484. A common beam of light is equivalent in all its effects to two beams vibrating in two rectangular planes. As two such beams cannot interfere, we cannot have the colors of the selenite in common light.

Rings surrounding the Axes of Crystals in Polarized Light.

485. A pencil of rays passing along the axis through Iceland spar suffers no division; but if inclined to the axis, however slightly, the pencil is divided into two, which vibrate in rectangular planes, and one of which is more retarded than the other.

486. If the incident light be polarized, on quitting the spar, oblique to the axis, it will be in a condition similar to the light emergent from the plates of gypsum already referred to. When two rectangular vibrations, passing through the same ether, are reduced to the same plane by the analyzer, interference occurs; the two rays either conspiring or opposing each other.

487. Whether they conspire or not depends upon the amount of relative retardation, and this again depends upon the thickness of the spar traversed by the two rays. If they conspire at a certain thickness they will also conspire at twice that thickness, thrice that thickness, etc. Those thicknesses at which the rays conspire are separated by others at which they oppose each other.

488. With a conical beam whose central ray passes *along* the axis, the effects are symmetrical all round the axis; and when the crystal, illuminated by such a ray, is examined by monochromatic polarized light, we have a series of bright and dark circles surrounding the axis.

489. When the light is red the circles are larger than when the light is blue; the smaller the wave-length the smaller are the circles. Hence, since the different colors are not superposed, when *white* light is employed instead of bands of alternate brightness and darkness we have a series of *iris-colored circles.*

When the polarizer and analyzer are crossed the system of bands is intersected by *a black cross*, whose arms are parallel to the planes of vibration in the polarizer and analyzer. Those rays, whose planes of vibration within the crystal coincide with the planes of either the polarizer or analyzer, *cannot get through either*, and their complete interception forms the two arms of the cross. Those rays whose planes of vibration enclose an angle of 45° with that of the polarizer or analyzer produce the greatest effect when they conspire. At this inclination the bright ring is at its maximum brilliancy, from which, right and left, it becomes more feeble, until it finally merges into the darkness of the cross.

490. A rotation of 90° produces here, as in other cases, the complementary phenomena: the black cross becomes white, and the circles change their tints to complementary ones.

491. In crystals possessing two optic axes a series of iris-colored bands surround both axes, each band forming a curve, which its discoverer, James Bernoulli, called a *lemniscata*.

Elliptic and Circular Polarization.

492. Two rays of light vibrating *at right angles to each other*, however the one system of vibrations may be retarded with reference to the other, cannot, as already stated, interfere so as to produce either an increase or a diminution of the light.

493. But though the *intensity* remains unchanged, the rays act upon each other. If one of them differs from the other by any exact number of semi-undulations, the two rays are compounded to a single *rectilinear* vibration. In all other cases the resultant vibration is *elliptical;* in one

particular case the ellipse in which the individual particles of ether move is converted into a *circle*. This occurs when one of the systems of waves is an exact quarter of an undulation behind the other; we have then *circular polarization*.

494. This compounding of ethereal vibration is mechanically the same as the compounding of the vibrations of an ordinary pendulum; or as the compounding of the vibrations of two rectangular tuning-forks by the method of Lissajous.*

495. Elliptic polarization is the *rule* and not the *exception*. It is particularly manifested in reflection from metals, and from transparent bodies which possess a high index of refraction. Jamin has detected it in light reflected from all bodies.

Rotary Polarization.

496. A polarized ray of monochromatic light, as already stated, suffers no change during its transmission through Iceland spar in the direction of the optic axis.

497. But if transmitted through *rock-crystal* (quartz) in the direction of the optic axis, its plane of vibration is turned by the crystal. Supposing the polarizer and analyzer of the polariscope to be crossed so as to produce perfect darkness before the crystal is introduced between them, on its introduction light will pass, and to quench the light the analyzer must be turned into a new position. The angle through which the analyzer is turned measures the *rotation of the plane of vibration*.

498. Some specimens of rock-crystal turn the plane of vibration to the right, and others to the left. The former are called *right-handed* and the latter *left-handed* crystals.

* See *Lectures on Sound*, 1st ed., p. 307.

Sir John Herschel connected this optical difference with a visible difference of crystalline form.

499. In the celebrated experiment of Faraday, with a bar of heavy glass, the plane of vibration was caused to rotate both by a magnet and an electric current; the direction of rotation bearing a constant relation to the polarity of the magnet and to the direction of the current.

500. The subject of rotary polarization was examined with great care and completeness by Biot, and he established certain laws regarding it, two of which may be enunciated here:

1. The amount of the rotation is proportional to the thickness of the plate of rock-crystal.

2. The rotation of the plane of vibration is different for the different rays of the spectrum, increasing with the refrangibility of the light.

Thus with a plate of rock-crystal one millimetre thick, he obtained the following rotations for the mean rays of the respective colors of the spectrum:

Red, 19°.	Green, 28°.	Indigo, 36°.
Orange, 21°.	Blue, 32°.	Violet, 41°.
Yellow, 23°.		

With a plate *two* millimetres in thickness the rotation for red is 38° and for violet 82°.

501. Since, then, the rays of different colors emerge from the rock-crystal vibrating in different planes, when such light falls upon the analyzer that color only whose plane of vibration coincides with that of the analyzer will be *transmitted*. By turning the analyzer we allow the other colors to pass in succession.

502. The phenomena of rotary polarization are produced by the interference of two *circularly-polarized pencils* of light, which are propagated along the axis with

unequal velocities, the one revolving from left to right, and the other revolving in the opposite direction.*

Conclusion.

I have endeavored in these lectures to bring before you the views at present entertained by all eminent scientific thinkers regarding the nature of light. I have endeavored to make as clear to you as possible that bold theory according to which space is filled with an elastic substance capable of transmitting the motions of light and heat. And consider how impossible it is to escape from this or some similar theory—to avoid ascribing to light, in space, *a material basis*. Solar light and heat require about eight minutes to travel from the sun to the earth. During this time the light and heat are detached from both. Enclose, in idea, a portion of the intervening space —say a cubic mile of it—occupied for a moment by light and heat. Ask yourselves what they are. The first inquiry toward a solution is, *What can they do?* We only know things by their *effects*. What, then, are the effects which this cubic mile of light and heat can produce? At the earth, where we can operate upon them, we find them capable of producing *motion*. We can lift weights with them; we can turn wheels with them; we can urge locomotives with them; we can fire projectiles with them. What other conclusion can you come to than that the light and heat which thus produce motion *are themselves motions?* †

One cubic mile of space, then, is for a measurable time the vehicle of motion. But is it in the human mind to imagine motion without at the same time imagining some-

---

* See Lloyd, *Wave Theory*, p. 199, etc.

† Sir William Thomson has attempted to calculate "the mechanical value of a cubic mile of sunlight."

thing moved? Certainly not. The very conception of motion necessarily includes that of a moving body. What, then, is the thing moved in the case of our cubic mile of sunlight? The undulatory theory replies that it is a substance of determinate mechanical properties, a body which may or may not be a form of ordinary matter, but to which, whether it is or not, we give the name of *ether*. Let us tolerate no vagueness here; for the greatest disservice that could be done to science—the surest way to give error a long lease of life—is to enshroud scientific theories in vagueness. The motion of the ether communicated to material substances throws them into motion. It is, therefore, itself *a material substance*, for we have no knowledge that in nature any thing but a material substance can throw other material substances into motion. Two modes of motion are possible to the ether. Either it is shot through space as a *projectile*, or it is the vehicle of *wave-motion*. The projectile theory, though enunciated by Newton, and supported by such men as Laplace, Biot, Brewster, and Malus, has hopelessly broken down. Wave-motion, then, of one kind or another, we must fall back upon. But how does the Wave Theory account for the phenomena? Throughout the greater part of these lectures we have been answering this question. The cases brought before you are *representative*. Thousands of facts might be cited in illustration of each of them, and not one of these facts is left unexplained by the undulatory theory. It accounts for all the phenomena of reflection; for all the phenomena of refraction, single and double; for all the phenomena of dispersion; for all the phenomena of diffraction; for the colors of thick plates and thin, as well as for the colors of all natural bodies. It accounts for all the phenomena of polarization; for all those wonderful affections, those chromatic splendors ex hibited by crystals in polarized light. Thousands of isolated facts might, as I have said, be ranged under each of these heads; the undulatory theory accounts for them all. It traces out illuminated paths through what would otherwise be the most hopeless jungle of phenomena in which human thought could be involved. This is why the foremost men of the age accept the ether not as a vague dream, but as a real entity—a substance endowed with *inertia*, and capable, in accordance with the established laws of motion, of imparting its thrill to other substances. If there is one conception more firmly fixed in modern scientific thought than another, it is that heat is a mode of motion. Ask yourselves how the vast amount of mechanical energy actually transmitted in the form of heat reaches the earth from the sun. *Matter* must be its vehicle, and the matter is according to theory the luminiferous ether.

Thomas Young never saw with his eyes the waves of sound; but he had the force of imagination to picture them and the intellect to investigate them. And he rose from the investigation of the unseen waves of air to that of the unseen waves of ether; his belief in the one being little, if at all, inferior to his belief in the other. One expression of his will illustrate the perfect definiteness of his ideas. To account for the aberration of light he thought it necessary to assume that the ether which encompasses the earth does not partake of the motion of our planet through space. His words are: "The ether passes through the solid mass of the earth as the wind passes through a grove of trees." This bold assumption has been shown to be unnecessary by Prof. Stokes, who proves that, by ascribing to the ether properties analogous to those of an elastic solid, aberration would be accounted for, without supposing the earth to be thus permeable. Stokes believes in the ether as firmly as Young did.

I may add, that one of the most refined experimenters in France, M. Fizeau, who is also a a member of the Institute, undertook to determine, some years ago, whether a moving body drags the ether along with it in its motion. His conclusion is, that *part of the ether* adheres to the molecules of the body, and is transferred along with them. This conclusion may or may not be correct; but the mere fact that such experiments were undertaken by such a man illustrates the distinctness with which this idea of an ether is held by the most eminent scientific workers of the age.

But while I have endeavored to place before you with the utmost possible clearness the basis of the undulatory theory, do I therefore wish to close your eyes against any evidence that may arise of its incorrectness? Far from it. You may say, and justly say, that a hundred years ago another theory was held by the most eminent men, and that, as the theory then held had to yield, the undulatory theory may have to yield also. This is perfectly logical. Just in the same way, a person in the time of Newton, or even in our own time, might reason thus: The great Ptolemy, and numbers of great men after him, believed that the earth was the centre of the solar system. Ptolemy's theory had to give way, and the theory of gravitation may, in its turn, have to give way also. This is just as logical as the former argument. The strength of the theory of gravitation rests on its competence to account for all the phenomena of the solar system; and how strong that theory is will be understood by those who have heard in this room Prof. Grant's lucid account of all that it explains. On a precisely similar basis rests the undulatory theory of light; only that the phenomena which it explains are far more varied and complex than the phenomena of gravitation. You regard, and justly so, the discovery of Neptune as a triumph of theory. Guided by it, Adams and Leverrier calculated the position of a planetary mass competent to produce the disturbances of Uranus. Leverrier communicated the result of his calculation to Galle, of Berlin; and that same night Galle pointed the telescope of the Berlin Observatory to the portion of the heavens indicated by Leverrier, and found there a planet 36,000 miles in diameter.

It so happens that the undulatory theory has also its Neptune. Fresnel had determined the mathematical expression for the wave-surface in crystals possessing two optic axes; but he did not appear to have an idea of any refraction in such crystals other than double refraction. While the subject was in this condition the late Sir William Hamilton, of Dublin, a profound mathematician, took it up, and proved the theory to lead to the conclusion that at four special points of the wave-surface the ray was divided not in *two* parts, but into an *infinite number of parts;* forming at those points a continuous *conical envelope* instead of two images. No human eye had ever seen this envelope when Sir William Hamilton inferred its existence. If the theory of gravitation be true, said Leverrier, in effect, to Dr. Galle, a planet ought to be there: if the theory of undulation be true, said Sir William Hamilton to Dr. Lloyd, my luminous envelope ought to be there. Lloyd took a crystal of Arragonite, and following with the most scrupulous exactness the indications of theory, discovered the envelope which had previously been an idea in the mind of the mathematician.

Whatever may be the strength which the theory of gravitation derives from the discovery of Neptune, it is matched by the strength which the undulatory theory derives from the discovery of *conical refraction.*

NOTE.

I would strongly recommend for perusal the essay on Light, published in Sir John Herschel's "Familiar Lectures on Scientific Subjects."

J. T.

NOTES

OF A COURSE OF SEVEN LECTURES ON

ELECTRICITY.

NOTES ON ELECTRICITY.

Voltaic Electricity: the Voltaic Battery.

1. IF two pieces of the same metal (pure zinc or pure platinum, for example) be immersed in water, which has been rendered sour by the addition of a little sulphuric acid, the acidulated water attacks neither.

The ordinary zinc of commerce being rendered impure by the admixture of other metals is attacked by the acid. It may, however, be enabled to withstand the acid by covering its surface with mercury. The zinc is dissolved by the mercury, detached from its impurities, and presented to the liquid. This process is called *amalgamation*.

2. If two pieces of two different metals (pure zinc and platinum, for example) be immersed in acidulated water, no sensible action occurs *as long as the metals do not touch each other;* but the moment they touch, and as long as they continue in contact, the zinc is attacked by the acidulated water and dissolves, while bubbles of gas rise from the surface of the platinum.

3. This gas when collected proves to have the specific gravity of hydrogen; like hydrogen it also burns in the air. The water, in fact, is decomposed by the touching metals; its oxygen unites with the zinc to form oxide of zinc, while its hydrogen escapes from the platinum.

4. If the two metals be only partially plunged into the acidulated water, it does not matter whether contact occurs *within* the liquid or *outside* of it. The effect in both cases is the decomposition of the water, the solution of the zinc, and the liberation of the hydrogen gas.

5. When the two partially immersed metals are connected outside the liquid by a long wire (say of copper) the effect is the same as when they touch directly. In both cases a *circuit* is said to be formed, consisting of the two metals and the liquid. In the case last mentioned the copper wire is said to complete the circuit.

For these experiments a strip of platinum and a strip of amalgamated zinc are employed. The liquid is placed in a glass cell with parallel sides, through which is sent a beam of light, and by means of a lens a magnified image of the cell and its two strips is cast upon a screen. The chemical action consequent upon touching the metals, or on completing the circuit with a wire, and its suspension when contact is interrupted, are then very plainly seen.

6. The wire is also said to be the vehicle of an *electric current* which flows round the circuit. It is also called a Voltaic current, because the action here described was discovered by the celebrated Italian philosopher Volta. These terms, however, convey to us, as yet, no meaning. Our sole business during the present lecture is to examine the wire which completes the circuit, and to determine wherein it differs from an ordinary wire.

7. And to enable ourselves to do this effectually, we shall employ an arrangement, or a combination, of zinc and platinum plates and acids, known as a voltaic battery. We shall subsequently analyze this battery, and determine what occurs within it. For the present, as aforesaid, we shall confine ourselves to the examination of the wire which completes the circuit outside the battery.

Electro-Magnetism: Elementary Phenomena.

8. Interrupting the circuit, and immersing the wire in iron filings, it shows no power of attraction over them. Establishing the circuit, on reimmersing the wire in the filings they cluster round it and cling to it. If the wire be raised out of the filings, they form an envelope round it. The moment, however, the circuit is interrupted, the filings fall.

9. If the wire be disconnected from the plates of platinum and zinc, and stretched under and parallel to a suspended bar magnet, no action is observed; but on making the wire, stretched beneath the magnet, form part of a voltaic circuit, the magnet is deflected from the magnetic meridian. This is Œrsted's discovery.

10. To the eye the wire, if tolerably thick, is unchanged by its connection with the zinc and platinum. But if for the thick copper wire a thin platinum wire be substituted it is sensibly heated, and may even be caused to glow brightly. The wire therefore must be the vehicle of some power or condition, which is competent to produce both magnetic and thermal phenomena.

11. If a naked wire, forming part of a voltaic circuit, be wound round a bar of iron, the power of which the wire was the vehicle is in great part transmitted to the iron which becomes part of the circuit.

12. But if the wire be overspun with cotton, or still better with silk, this transmission of the power from the wire to the iron bar is prevented. The wire may then be coiled round the bar while the power is compelled to pass in succession through all the convolutions of the wire. Here the iron bar is not at all in the circuit.

13. But though not in the circuit it is powerfully excited by the surrounding wire. Every convolution of the

wire evokes a certain amount of *magnetism* in the bar; and by rendering the convolutions sufficiently numerous, a magnet of enormous strength may be thus generated. This is Sturgeon's application of Arago's discovery.

14. Such a magnet is called an Electro-magnet to distinguish it from ordinary permanent steel magnets. When the circuit is broken the power of the electro-magnet ceases. It then falls from its highly-excited condition to the condition of ordinary iron.

15. For electro-magnetic purposes the covered wire is usually coiled round a hollow reel, several layers of coil being sometimes superposed upon each other. In this condition the reel is called an *electro-magnetic helix*. The iron bar to be magnetized is placed within the helix, forming its *core*. The electro-magnet may be either straight, shaped like a horseshoe, or it may be caused to assume other forms.

16. The smooth bar of iron placed across the ends, or poles, of a horseshoe magnet, is sometimes called a *keeper*, sometimes an *armature*, and sometimes a *sub-magnet*.

17. It is not necessary that the convolutions of the helix should be close to the core. A hoop, for example, a yard in diameter, round which covered wire is coiled, magnetizes an iron bar placed across it at its centre. The magnetized body is here nearly 18 inches from the magnetizing coil. How is the power transmitted from the one to the other? Is it an action at a distance, or does it require a medium for its propagation? I do not know. The question at present profoundly interests investigators.

18. If a covered wire forming part of a voltaic circuit be coiled round an iron bar near one of its ends, there is a propagation of the excitement along the bar toward the distant end. As the coils augment in number the attrac-

tive power of the distant end increases. On undoing the coils the magnetism gradually falls. The process resembles more or less the conduction of heat. The augmentation of the coils answering to the increasing of the temperature, and the undoing of the coils answering to the cooling of the end of the bar.

Electro-Magnetic Engines.

19. When the end of a cylinder of iron is partially introduced into an electro-magnetic helix, on completing the circuit a force of suction is exerted upon it tending to draw it into the helix. Page turned this force to account in the construction of an electro-magnetic engine.

Hollow iron cylinders, which pass freely into the helix, are employed for this experiment. The end only of the hollow cylinder being introduced, when the circuit is completed the cylinder is suddenly and strongly sucked in.

20. Others have turned to account mechanically the attraction exerted by electro-magnetic cores on bars of iron. The distinguished electro-mechanician Froment produced rotatory motion in this way. A series of electro-magnets are so ranged that their poles lie facing each other along the circumference of a circle; and a series of transverse bars of iron are so connected together as to be able to approach the poles in succession, and rotate as a system. When the circuit is established, these bars are attracted, motion being thus imparted to the system. The bars on arriving at the poles which attract them suddenly cease to be attracted; the magnetism being temporarily suspended to allow each bar to pass forward, with the velocity impressed upon it, to the next pair of attracting poles. On reaching these the magnetism is again temporarily suspended. Thus the bars *are never pulled*

back; and in this way a continuous motion of rotation is maintained.

21. This rotatory motion can be applied in various ways; it may, for example, be caused to pump water, to saw wood, or to drive piles.

One of Froment's electro-magnetic engines, and its application to pumping and pile-driving, is employed to illustrate this.

Physical Effects of Magnetization.

22. Sound is one of the physical effects which accompany sudden magnetization and sudden demagnetization. An ear placed close to an iron core hears a clink the moment the circuit is established round it. A clink is also heard when the circuit is broken. This is Page's discovery. Employing a contact-breaker (in a distant room to abolish its noise) the coil may be magnetized and demagnetized in quick succession; the sounds then produced may be heard by several hundreds at once.

A poker of good soft iron placed within an electro-magnetic helix, and with its two ends supported on wooden trays, produces a very good effect. The sound may be rendered musical.

23. When an iron bar is magnetized its volume is unchanged, but its shape is altered. It lengthens in the direction of magnetization. This is Joule's discovery.

24. Joule employed a system of levers to augment the effect, and a microscope to observe the elongation thus augmented. Our method is this: The iron bar is magnetized by an electro-magnetic helix which surrounds it. Its elongation is first augmented fiftyfold by means of a lever; and this motion is applied to turn the axis of a rotating mirror. From the mirror is reflected a long beam of light, which forms an index without weight. The re-

flected beam may be caused to print a circle of light upon a white screen, and this circle when the bar is magnetized, suffers a displacement due to the elongation of the bar. This displacement may amount to a foot or more.

What is the cause of this elongation? The discussion of this question requires some preliminary knowledge.

25. If a sheet of paper or a square of glass be placed over a magnet, iron filings scattered on the paper or on the glass arrange themselves in lines, which Faraday called Lines of Force. Along these lines the filings set their longest dimensions, and they also attach themselves end to end. A little bar of iron, or a small magnetic needle, freely suspended, sets itself also along these lines of force.

The formation and modifications of the magnetic curves, or lines of force, are shown in this lecture by means of small magnets held between plates of glass and strongly illuminated. Magnified images of the curves are thrown upon a screen about 40 feet distant. The shifting of the curves by the tapping of the glass is plainly visible.

26. We may regard a bar of iron as made up of particles united by the force of cohesion, but still to some extent distinct. When iron is broken we see crystalline facets on the surface of fracture. In fact, the bar is composed of minute crystals of irregular shape. These, when the bar is magnetized, try to set their longest dimensions parallel to the direction of magnetization, that is to say, in the direction of the bar itself. They succeed in this effort to some slight extent, and thus produce the minute and temporary lengthening of the bar. This is the explanation of De la Rive. It is, I think, as true as it is acute.

27. Magnetic oxide of iron may be suspended as a powder in water contained in a cylindrical vessel with flat

glass ends. Let the vessel be surrounded by a coil of covered wire. Looking at a candle through the muddy liquid, and making the coil part of a voltaic circuit, the candle brightens at the moment the circuit is made. Breaking the circuit, dimness again supervenes. This is due to an arrangement of the particles of suspended oxide similar to that of the iron filings. They set their longest dimensions parallel to the beam of light, and thus obstruct its passage less. They also attach themselves end to end, and form lines like the lines of filings. This beautiful experiment is due to Grove.

Projecting a magnified image of the end of the cylindrical cell on a screen, and sending through it the beam of the electric lamp whenever the circuit is established, an illuminated disk, 2 or 3 feet in diameter, flashes out upon the screen.

Character of Magnetic Force.

It is necessary to our further progress to have clear and definite ideas as to the character of the magnetic force.

28. The magnetic power of a magnet, or of a magnetic needle, though really distributed throughout its mass, appears to be concentrated at two points near the ends. These points are called the *poles* of the magnet or needle.

29. The magnetic power of the earth is doubtless also distributed through the mass of the earth, but a concentration similar to that just noticed endows the earth also with magnetic poles.

30. The action of the earth upon a magnetic needle is this: the north terrestrial pole repels one end of the needle and attracts the other; the south magnetic pole also attracts one end of the needle and repels the other.

CHARACTER OF MAGNETIC FORCE. 139

But the end attracted by the north terrestrial pole is repelled by the south, while the end attracted by the south is repelled by the north.

31. Thus to each terrestrial magnetic pole the needle presents two ends which are differently endowed. Two opposite kinds of magnetism may be supposed to be concentrated at the two ends. In this *doubleness* of the magnetic force consists what is called *magnetic polarity*.

32. Each of the two distinct kinds of magnetism may be regarded as self-repellent. North repels north, and south repels south. But different kinds of magnetism are mutually attractive; south attracts north, and north attracts south.

33. When a magnet, or a magnetic needle, is suspended with the line joining its poles *oblique* to the magnetic meridian, the earth's action on the needle resolves itself into what in mechanics is called "a couple," tending to turn the needle into the magnetic meridian.

34. When the needle is in the meridian, the two forces which constitute the couple are opposite and equal. The tendency to produce rotation then ceases; the needle is in its position of equilibrium.

35. When the forces are equal and opposite they must neutralize each other; no *motion of translation* of the needle being, therefore, possible. Thus, when the needle is caused to swim on water, or on mercury, it does not move toward either of the terrestrial magnetic poles.

36. One pole of a bar magnet repels the one end and attracts the other end of a magnetic needle. At the other pole of the magnet the attraction and repulsion are reversed. In the middle of the magnet is the *magnetic equator*, where neither end of the needle is attracted or repelled.

Magnetism of Helix: Strength of Electro-Magnets.

37. An electro-magnetic helix, even without a core of iron, behaves exactly like a magnet. It attracts iron. Its two ends, moreover, are opposite poles, and between them is a magnetic equator. When, however, a core is placed within the helix, the magnetism of the combined system is far more intense than that of the helix alone.

38. The strength of a magnet is measured by its power to deflect a magnetic needle from its meridian; the magnetic strength of a helix alone, and of a helix and core combined, are similarly determined.

39. To obtain the magnetic strength of the core alone, we first determine the strength of the helix alone, then that of the helix and core combined; subtracting the former strength from the latter, we obtain the magnetic strength of the core.

40. If the cores be thick and formed of good iron, the magnetic strength of the core is exactly proportional to that of the helix. A helix of double power will produce an electro-magnet of double strength; a helix of treble power, an electro-magnet of treble strength, and so on. Thus by varying the strength of the helix we vary in like degree the strength of the iron core within it.

Electro-Magnetic Attractions: Law of Squares.

41. And here an important point arises. When we allow a core of double power to act upon a piece of good iron, nearly but not quite in contact with the core, the attraction of the iron is not doubled, but quadrupled. If the core be of treble power, the attraction is not only trebled, but it increases ninefold. If the magnetic strength of the core be quadrupled, the attraction of the iron is augmented sixteenfold. In fact, the attraction is

proportional, not to the strength simply, but to the strength multiplied by itself, or to the *square of the strength* of the electro-magnet.

We must be very clear as to the cause of this action, and must, therefore, contrast for a moment the magnetic action of hard steel with that of soft iron.

42. Soft iron is easily magnetized, but it loses its magnetism when the magnetizing force is withdrawn. Steel is magnetized with difficulty, but it retains its magnetism even after the withdrawal of the magnetizing magnet.

43. This obstinacy on the part of steel in declining to accept the magnetic state, and this retentiveness on the part of steel when the magnetic condition has been once imposed upon it, are called *coercive force*. It is not a happy term, but it is the one employed.

44. Supposing a piece of magnetized steel to possess a coercive force so high as to resist further magnetization, its attraction by an electro-magnet would be directly proportional, not to the square of the strength, but simply to the strength of the electro-magnet.

45. Why, then, does the iron follow the law of the square of the strength? It is because the magnetic condition of the iron is not constant, but rises with the strength of the magnet. When the magnetism of the core is doubled, the magnetism of the iron is also doubled; when the magnetism of the core is trebled, the magnetism of the iron is trebled. The resultant attraction is found by multiplying the magnetism of the iron by the magnetism of the core, and this product is the expression of the law of squares just referred to.

46. To make the matter clearer, let us figure the magnetism of the core as due to particles of magnetism, which are introduced into the core in gradually-increasing numbers. Let us start with a core possessing one magnetic

particle, and let it act upon a piece of hard steel also possessing one magnetic particle; the resulting attraction will be unity or 1. Let two particles be now thrown into the core: the steel in virtue of its coercive force remains unchanged, but its particle being now pulled by two particles instead of one, the resulting attraction will be 2. If three particles of magnetism be thrown into the core, all of them pulling at the single particle of the steel will produce a treble attraction, and so on.

47. Now let us start with a core possessing, as before, a single particle of magnetism, and with a piece of iron also possessing a single particle generated by the core; the attraction, as before, is here unity. On introducing two particles into the core, they generate immediately two particles in the iron. But two particles each pulled by twice the force first exerted, makes the attraction four times what it was at the outset.

It is to be remembered that every particle is attracted as if the other particles were absent.

48. In like manner, if three particles be thrown into the core, three particles are also generated in the iron. Each of these iron magnetic particles is pulled by the three particles of the electro-magnet; that is to say, each of the iron particles is pulled with three times the primitive force. But there are three particles so pulled; hence the attraction is nine times what it was at the outset.

49. Let us compare this action for a moment with that of gravity. Two masses of matter attract each other with a force which we shall take as our unit. If the one mass be doubled, the attraction is doubled; if both masses be doubled, the attraction is increased fourfold. If one mass be trebled, the attraction is trebled; if both masses be trebled, the attraction is increased ninefold. When, therefore, both the masses are doubled and trebled, we

have the law of squares. Now, it is this doubling and trebling, *in both cases*, of the thing which causes magnetic attraction, which causes it to follow the same law.

Inference from Law of Squares: Theoretic Notions.

50. Why do I lead you through these considerations? Simply to make clear to you, that if "the law of squares" here developed show itself in the action of a magnet upon matter, we may infallibly infer that the condition of that matter is not *a constant condition;* but that it rises and falls with the condition of the magnet. Matter thus affected is said to be magnetized by influence or by induction. It is attracted or repelled (for we shall come immediately to the repulsion of matter by a magnet) in virtue of some condition into which it is temporarily thrown by the influencing magnet.

51. What then is the thing that causes magnetic attraction? The human mind has long striven to realize it. Thales (600 B.C.) thought that the magnet possessed a soul. Cornelius Gemma in 1535 supposed invisible lines to stretch from the magnet to the attracted body, a conception which reminds us of Faraday's Lines of Force. Others thought the iron the natural nutriment of the magnet. Descartes embraced magnetic phenomena in his celebrated theory of vortices, and in our day Clerk Maxwell has worked in this direction. Æpinus assumed the existence of a magnetic fluid. Coulomb assumed the existence of two fluids, each self-repellent, but mutually attractive. Ampère deemed a magnet an assemblage of minute electric currents, which circulated round the atoms of the magnetized body. These conceptions are sometimes exceedingly useful as a means of connection and classification, even when we do not believe them true. William Thomson deduces magnetic phenomena from

"imaginary magnetic matter," thus giving the mind the conception while distinctly releasing it from belief. The real origin of magnetism is yet to be revealed.

Diamagnetism: Magne-Crystallic Action.

52. Brugmans, in 1778, first observed the repulsion of bismuth by a magnet. In 1827 Le Baillif described the repulsion of antimony. Saigey, Seebeck, and Becquerel, also observed certain actions of the kind.

53. In 1845 Faraday generalized these observations by demonstrating the magnetic condition of all matter. He showed that bodies divided themselves into two great classes, the one attracted, the other repelled by the poles of a magnet.

54. To the force producing this repulsion, Faraday gave the name of Diamagnetism.

What is the nature of this force? Is it inherent and constant, or is it induced?

55. The repulsion of diamagnetic bodies follows accurately the law of squares above developed. A double force produces a quadruple repulsion; a treble force produces a ninefold repulsion, and so on.

56. Hence we may infer, with certainty, that the condition of diamagnetic bodies in virtue of which they are repelled by a magnet, is a condition induced by the magnet, rising and falling as the strength of the magnet rises and falls.

57. The force of diamagnetism is vastly feebler than that of ordinary magnetism. Of all diamagnetic substances, for example, bismuth is the most strongly repelled; but its repulsion is almost incomparably less than the attraction of iron. According to Weber, the magnetism of a thin bar of iron exceeds the diamagnetism of an equal mass of bismuth about two and a half million times.

58. Diamagnetic bodies under magnetic excitement exhibit a polarity the reverse of that of magnetic bodies. In all cases, whether we operate with helices or with magnets, or with helices and magnets combined, the actions of magnetic and diamagnetic bodies are antithetical.

59. An iron statue standing erect on the earth's surface is converted into a magnet by the earth's magnetism; a marble statue, or a man standing erect, is converted by the same force into a diamagnet; for marble is diamagnetic, and so are all the tissues and all the solids and fluids of the human body. The poles of the man are those of the iron statue reversed.

60. Organic bodies, and most crystals, are magnetized with different degrees of intensity in different directions. They are endowed with *axes* of magnetic induction.

61. Thus in the case of Iceland spar (carbonate of lime), the *repulsion* along the axis is a maximum. In the case of carbonate of iron, a crystal of the same shape and structure as carbonate of lime, the *attraction* along the axis is a maximum.

62. The position assumed by a crystal when suspended between the poles of a magnet, depends on its magnetic axes.

Frictional Electricity: Attraction and Repulsion: Conduction and Insulation.

63. By the friction of a woollen cloth amber is endowed with the power of attracting light bodies. This substance was called Electron by the Greeks; hence the name Electricity was applied to the power of attraction exhibited by amber. This attraction remained an isolated fact for more than 2,000 years.

64. In the year 1600 Dr. Gilbert of Colchester, physi-

cian to Queen Elizabeth, showed that the power of attraction was shared by many other substances. Dry glass, for example, when rubbed by silk, and dry sealing-wax when rubbed by flannel, exhibit this attractive power. When they do so they are said to be *electrified*.

65. An electrified body attracts and is attracted by all kinds of unelectrified matter; but *repulsion* may also come into play. Thus, rubbed glass *repels* rubbed glass, and rubbed sealing-wax repels rubbed sealing-wax; while rubbed glass attracts rubbed sealing-wax, and rubbed sealing-wax attracts rubbed glass.

66. Hence the notion of *two kinds* of electricity: one proper to vitreous bodies, and therefore called vitreous electricity; the other proper to resinous bodies, and therefore called resinous electricity.

67. These terms are improper; because by employing suitable rubbers we can obtain the electricity of sealing-wax from glass, and the electricity of glass from sealing-wax. We now use the term *positive* electricity to denote that developed on glass by the friction of silk; and *negative* electricity to denote that developed on sealing-wax by the friction of flannel.

68. Bodies endowed with the same electricity repel each other, while bodies endowed with opposite electricities attract each other. This is the fundamental law of electric action.

69. The rubber and the body rubbed are always endowed with opposite electricities. They always attract each other. The work done in overcoming this attraction appears as heat in the electric spark.

70. To find the kind of electricity with which a body is endowed we must ascertain, by trial, the electricity by which the body is *repelled*. This, we may be sure, is the electricity of the body. *Attraction* does not furnish a safe test, because unelectrified bodies are attracted.

71 Some substances possess in a very high degree the capacity of transmitting the electric power, or condition; others possess in a high degree the capacity of intercepting it. The former bodies are called *conductors*, the latter bodies, *insulators*.

72. The insulators were formerly called *electrics*, because they could be electrified by friction *when held in the hand*. The conductors were called *non-electrics*, because they could not be so electrified. The division is improper, because if a conductor be insulated it can readily be electrified. To keep it electrified an insulator must be introduced between it and the earth.

Theories of Electricity: Electric Fluids.

73. What is electricity? Why should it adhere so tenaciously to some substances, and flow so freely through or along others? The human mind has made many attempts to imagine the inner cause of electric action, and it still continues to make such attempts. Formerly it was thought that magnetism and electricity, as well as light and heat, were all the work of "imponderable matter," associated with the ordinary matter. In the case of light and heat, this conception has undergone profound modification; and we seem to see clearly the mechanical cause of both. But no similar clearness has as yet been attained with regard to electricity, though a strong presumption exists that our notions of it are destined soon to undergo a modification equally profound.

74. Meanwhile we may employ the provisional conception furnished by the *theory of electric fluids*. It will enable us to classify our facts, though it is not to be regarded as demonstrated.

75. According to this theory, electrical attractions and repulsions arise from two invisible fluids, each self-re-

pulsive but both mutually attractive. The fluids are supposed to be mixed together to form a compound neutral fluid in unelectrified bodies.

76. The act of electrification, by friction, consists in the forcible separation of the two fluids, one of which is diffused over the rubber, and the other over the body rubbed. But they may also be separated in another way now to be illustrated.

Electric Induction: the Condenser: the Electrophorus.

77. If an electrified body be brought near an insulated unelectrified conductor, but not into contact with it, the electrified body will decompose the compound fluid of the conductor; attracting one of its constituents and repelling the other. When the electrified body is withdrawn, the separated fluids reunite and neutralize each other.

78. This forcible separation of the two fluids of a neutral conductor, by the mere proximity of an electrified body, is called *electric induction.* Bodies in this state are also said to be electrified by *influence.* Neutral bodies are attracted because they are first excited by induction.

79. When an insulated conductor is acted on by an electrified body, its repelled electricity is free, but its attracted electricity is held captive by the inducing electrified body. Connecting the conductor for a moment with the earth, its free electricity escapes; and then, on the removal of the electrified inducing body, the captive electricity is liberated and diffused over the surface of the conductor.

80. Thus by the mere proximity of the electrified body, and without establishing contact between it and the neutral conductor, we can charge the latter *with the opposite electricity.*

81. Two sheets of tin-foil (conductors) being separated

from each other by a sheet of glass (an insulator), if one sheet have electricity imparted to it, it will act through the glass, and decompose the neutral electricity of the opposite sheet attracting the one constituent and repelling the other.

82. If the second sheet be connected with the earth the repelled electricity will flow away, and we shall have two mutually attractive layers of electricity separated from each other by the glass.

83. If the one sheet of tin-foil be united with the other by a conductor, the two opposite electricities will flow together; the tin-foil is then said to be discharged. This discharge usually assumes the form of a spark.

84. If the surface of a cake of resin, or of a sheet of vulcanized india-rubber be electrified, a plate of metal laid upon it will have its neutral fluid decomposed; its positive fluid being attracted and its negative repelled. On touching the metal plate its free (repelled) electricity flows to the earth; and now if the plate be raised by an insulating handle, it will appear charged with positive electricity. This is the principle of the *Electrophorus*.

The Electric Machine: the Leyden-jar.

85. An Electric Machine consists of two parts: the insulator, which is excited by friction, and the prime conductor.

86. The first electric machine consisted of a ball of sulphur, which was rubbed against the hand. It was invented by Otto von Guericke, burgomaster of Magdeburg, in the year 1671. A sphere of glass was afterward introduced, then a cylinder of glass, and finally a round glass plate, which was rubbed with dry silk.

87. The prime conductor is thus charged: When the glass plate is turned by a handle it passes between the silk

rubbers and is positively electrified. The electrified glass then acts by induction upon the prime conductor, attracting the negative electricity and repelling its positive. The conductor is furnished with points, from which the negative electricity streams out against the excited glass. Thus the prime conductor is charged, not by directly communicating to it positive electricity, but by robbing it of its negative, the positive remaining behind.

88. The arrangement mentioned in Note 81 is virtually a Leyden-jar. Were the plate of glass there referred to moulded into the shape of a jar, one sheet of foil would cover its interior and the other its exterior. When the jar is connected with an electric machine, its charged interior coating acts by induction across the glass on the exterior coating, attracting the opposite and repelling the similar electricity.

89. In the experiment which led to the discovery of the Leyden-jar *the hand of the experimentalist* served as the outer coating.

90. The escape of the repelled electricity of the outer coating to the earth leaves the captive electricity exposed solely to the attraction of that within the jar, and enables the jar to take a strong charge.

The Electric Current.

91. When the outer and the inner coatings are connected by a conductor, *an electric current* passes from the one to the other.

92. The current starts at the same instant from the inner and outer coatings; the *middle* point of the conductor being reached last by the current. This indicates that there are *two* currents which start at the same moment from the inner and outer coatings.

THE ELECTRIC DISCHARGE.

93. It is agreed to call the direction in which the *positive* electricity flows *the direction of the current.*

The Electric Discharge: Thunder and Lightning.

94. When an electric current encounters resistance in its passage, heat is developed: this heat is sometimes so intense as to reduce metals to a state of vapor.

95. When a body is intensely electrified, it will discharge its electricity to an unelectrified body across an interval of air in the form of an electric spark. Two bodies oppositely electrified discharge to each other in the same way.

96. When two oppositely electrified clouds discharge toward each other, the track of the lightning marks the course of an electric current, and the sound of the thunder is the sound of an electric spark.

97. An electrified cloud, if it come near the earth, may discharge its electricity to the earth in the same way.

98. If the body through which the atmospheric electricity passes be a good conductor, and of sufficient size, no harm is done; but the *resistance* offered by trees, houses, and animals, to the passage of the electricity usually causes their destruction.

99. The nervous system requires a certain interval of time to become conscious of pain. The time of an electric discharge is but a small fraction of this interval; hence as a sentient apparatus the nervous system is destroyed before consciousness can set in. If this be true—and there are the strongest grounds for believing it to be true—death from lightning must be painless.

100. When an electrified cloud passes over a pointed lightning-conductor, the opposite electricity of the earth is discharged from the point of the conductor against the

cloud. The cloud is thus neutralized, and, in general, without producing thunder.

101. The duration of an electric spark amounts only to an extremely small fraction of a second. On this account, when moving bodies are suddenly illuminated by the spark from a Leyden-jar, they appear to *rest* for a short interval in the position which they occupied when the flash fell upon them. A moving cannon-ball illuminated by a flash of lightning appears to stand still about one-eighth of a second, this being about the interval during which an impression, once made, persists upon the retina.

102. The unretarded electric spark will scatter gunpowder, but will not ignite it. To produce ignition it is necessary to retard the discharge by sending it through a wet string.

Electric Density: Action of Points.

103. If we double the quantity of electricity imparted to the same conductor, the density of the electricity is said to be doubled; if we treble the quantity, the density is said to be trebled; and so on.

104. On a *sphere* the density of the electricity is the same at all points of its surface; on a *plate* the density is greatest at the edges; and on an elongated conductor the density is greatest at the ends.

105. When the conductor ends in a sharp point the electric density at the point is so great that the electricity discharges itself into the air.

106. The air thus electrified is self-repellent, and is also repelled by the point, the so-called "electric wind" being produced.

107. By causing an electric wind to issue from opposite points of a light body, the reaction of the two winds

RELATION OF VOLTAIC TO FRICTIONAL ELECTRICITY.

may make the body to float in stable equilibrium in the air.

Relation of Voltaic to Frictional Electricity.

108. The outer ends of two pieces of zinc and platinum, partially immersed in acidulated water, are in opposite electrical conditions. The free platinum end shows positive electricity, while the free zinc end shows negative electricity.

109. When both plates are united by a wire, the positive flows along the wire toward the negative, and the negative toward the positive. But, as mentioned in Note 93, it is agreed to call the direction in which the positive electricity flows *the direction of the current.*

110. The force which urges this current forward (the electro-motive force) is enormously less than that which urges forward a current of frictional electricity. The consequence is, that the latter is able to surmount resistances which are totally unsurmountable by the former.

111. But by linking cells together we cause the voltaic current to approach more and more to the character of the frictional current. It requires, however, a battery of more than a thousand cells to make the current from a voltaic battery jump over an interval of air $\frac{1}{1000}$th of an inch in length. An electric machine of moderate power, and furnished with a suitable conductor, is competent to urge its current across an interval ten thousand times as great as this.

112. The electric spark passes through air by the agency of the particles of the conductor from which it springs, and which are carried forward by the discharge.

113. But measured by other standards the frictional current is almost incomparably more feeble than the voltaic current. For example: it is not without special

arrangements for multiplying the effect that the current from a large electrical machine is enabled to deflect a magnetic needle.

114. Faraday immersed two wires, the one of zinc and the other of platinum, each $\frac{1}{18}$th of an inch in diameter, in a cell of acidulated water. The depth of immersion was only $\frac{5}{8}$ths of an inch, and the time of immersion only $\frac{3}{20}$ths of a second. Still he found that the electricity generated by this small apparatus, in this brief time, produced a distinctly greater effect upon a magnetic needle than 28 turns of the large electric machine of the Royal Institution.

115. A cubic inch of air, if compressed with sufficient power, may be able to rupture a very rigid envelope; while a cubic yard of air, if not so compressed, may exert but a feeble pressure upon the surfaces which bound it. Now the electricity of the machine is in a condition analogous to the compressed air. Its density, or, as it is sometimes called, its intensity, or tension, is great. The electricity from the voltaic battery, on the other hand, resembles the uncompressed air. It exceeds enormously in *quantity* that from the machine; but it falls enormously below it in intensity.

116. The deflection of a magnetic needle and other actions of the voltaic current depend solely upon quantity, hence the vast superiority of the voltaic current in producing such deflection.

117. Faraday found the quantity of electricity disengaged by the decomposition of a single grain of water in a voltaic cell (see Note 5) to be equal to that liberated in 800,000 discharges of the great Leyden battery of the Royal Institution. This, if concentrated in a single discharge, would be equal to a great flash of lightning. He also estimated the quantity of electricity liberated by the

chemical action of a single grain of water on four grains of zinc to be equal in quantity to that of a powerful thunder-storm.

118. Weber and Kohlrausch have found that the quantity of electricity associated with one milligramme of hydrogen in water, if diffused over a cloud 1,000 mètres above the earth, would exert upon an equal quantity of the opposite electricity at the earth's surface an attractive force of 2,268,000 kilogrammes.*

Historic Jottings, concerning Conduction and the Leyden-jar.

119. In 1729, Stephen Grey, pensioner of the Charter House, discovered electric conduction. Connecting an end of a wire 700 feet long with a glass tube and supporting the wire on loops of silk, he found that on rubbing the tube the distant end of his wire became electrified and attracted light bodies. He also found that a wire loop did not answer as a support, as the electricity escaped through it; hence arose the division of bodies into conductors and insulators. Grey's observations were written down by the secretary of the Royal Society the day before his death.

120. In October, 1745, Von Kleist, a bishop of Cammin, in Pomerania, charged with electricity a flask containing sometimes mercury, sometimes alcohol. Through a cork in the neck of the flask passed an iron nail, which was brought into contact with the conductor of an electrical machine. On touching the nail Von Kleist experienced a violent shock.

121. In January, 1746, Cunæus of Leyden received also a shock, and his experiment was repeated by Allamand

* The mètre is a yard and one-eleventh in length; the milligramme is $\frac{1}{68}$th of a grain; the kilogramme is 2 lbs. $3\frac{1}{4}$ oz.

and Musschenbroek. A wire passed from the conductor of the machine into a flask filled with water. Musschenbroek held the flask in the right hand, the machine was turned, and then with the left hand he drew a spark from the conductor. The shock received was, according to Musschenbroek so terrible, that he declared he would not receive a second for the crown of France. Musschenbroek observed that it was only the person who held the flask in his hand that felt the shock. Kleist failed to recognize this condition.

122. In Germany the jar is sometimes called Kleist's jar, but more commonly, because of the failure just referred to, the Leyden-jar. The theory of it, and other similar apparatus, was given by Franklin in September, 1747. (See Notes 81, 88, 89, 90.)

123. In 1747, Dr. Watson, Bishop of Llandaff, sent the discharge from a Leyden-jar through 2,800 feet of wire, and through the same distance of earth. Subsequently, in the same year, he sent the discharge through 10,600 feet of wire, supported by insulators of baked wood. The experiment was made on Shooter's Hill.

124. In 1748 similar experiments were made by Franklin across the Schuylkill, and by De Luc across the Lake of Geneva.

Historic Jottings, concerning the Electric Telegraph.

125. The first proposal of an electric telegraph was made by an anonymous contributor to the *Scot's Magazine* for 1753. Various attempts to apply frictional electricity for this purpose were subsequently made. They culminated in the exceedingly ingenious arrangement of Mr. (now Sir Francis) Ronalds, published in 1823.

126. The voltaic pile was described by Volta in a letter to Sir Joseph Banks, written from Como in 1800.

127. Immediately afterward Nicholson and Carlisle discovered the decomposition of water by the voltaic current.

128. In 1808 Sömmering proposed a system of telegraphy based on the discovery of Nicholson and Carlisle. A similar system was proposed about the same time by Prof. Coxe, of Pennsylvania.

129. In 1820 Œrsted discovered the deflection of a magnetic needle by an electric current.*

130. The idea of employing the deflection of the needle for telegraphic purposes occurred to the celebrated French mathematician, La Place; the problem was partly worked out by Ampère, and still further advanced by Ritchie, Professor of Natural Philosophy in the Royal Institution.

131. In 1832 Baron Schilling constructed models of a telegraphic apparatus which were exhibited before the Emperors Alexander and Nicholas.

132. In 1833 Gauss and Weber established an electric telegraph between the Physical Cabinet and the Astronomical and Magnetic Observatories of Göttingen, embracing a distance of nearly 10,000 feet. Faraday's electricity instead of Volta's was employed by Gauss and Weber.

133. Steinheil was requested by Gauss to pursue the subject. To the telegraph he made many highly-impor-

* In his exceedingly useful little book on the Telegraph, published in Weale's "Rudimentary Series," Mr. Robert Sabine quotes the following remarkable passage from a work on magnetism, published in Paris, by Prof. Izarn, in 1804: "D'après les observations de Romagnési, physicien de Trente, l'aiguille déjà aimantée, et que l'on soumet ainsi au courant galvanique, éprouve une déclinaison; et d'après celles de J. Majon, savant chémiste de Gônes, les aiguilles non-aimentées acquièrent par ce moyen une sorte de polarité magnétique." The work containing this passage was lent to Mr. Sabine by Mr. Latimer Clark.

taut contributions and suggestions. In 1837 he had established a system of wires about 40,000 feet in length, connecting various points in the city of Munich and its neighborhood. The most considerable discovery of Steinheil, and, indeed, one of the most practically important hitherto made in connection with telegraphy, is that the "return wire" between two stations might be dispensed with, and the earth employed in its stead.

134. In 1834 Wheatstone, by means of a rotating mirror, made his celebrated experiments on the velocity of electricity. In the following year he exhibited one of Baron Schilling's telegraphs in his lectures at King's College.

135. In 1836 Mr. William Fothergill Cooke saw in the lectures of Prof. Muncke, at Heidelberg, the performance of a similar instrument. Struck by its obvious practical importance, he devised a system of telegraphy, and, in partnership with Wheatstone, dating from June, 1837, succeeded in introducing the telegraphic system into England.

136. From 1832 to 1836 Morse sought to apply chemical decomposition by the electric current to telegraphic purposes; he abandoned this for his electro-magnetic system devised in 1836. This method consists in stamping, by means of the attraction of an electro-magnet, dots and lines upon a slip of paper caused to move by proper mechanism over the circumference of a wheel.

137. In 1850 the first submarine cable was laid by Mr. Brett between Dover and Calais. It survived only a day. In 1851 another cable was laid down, which proved successful.

138. On the 5th of August, 1858, the submergence of the first Atlantic cable was completed, and messages were sent between England and America. The cable ceased

to act on the 4th of September, or about a month after its submersion.

139. In 1865 the second Atlantic cable was laid and lost. In 1866 a cable was successfully laid, and in the same year the cable of 1865 was recovered. Messages are now sent between England and America at the rate of fourteen words a minute.

Phenomena observed in Telegraph-Cables.

140. Davy showed ("Elements of Chemical Philosophy," 1812, p. 154) that a Leyden-battery could be charged with voltaic electricity.*

---

* Davy thus describes the celebrated battery with which he made this experiment. The spirit to which the battery owed its birth has not diminished among the members of the Royal Institution: "The most powerful combination that exists in which number of alternations is combined with extent of surface, is that constructed by the subscriptions of a few zealous cultivators and patrons of science, in the laboratory of the Royal Institution (in 1808). It consists of two hundred instruments, connected together in regular order, each composed of ten double plates arranged in cells of porcelain, and containing in each plate thirty-two square inches; so that the whole number of double plates is 2,000, and the whole surface 128,000 square inches. This battery, when the cells were filled with 60 parts of water mixed with one part of nitric acid, and one part of sulphuric acid, afforded a series of brilliant and impressive effects. When pieces of charcoal about an inch long and one-sixth of an inch in diameter, were brought near each other (within the thirtieth or fortieth part of an inch) a bright spark was produced, and more than half the volume of the charcoal became ignited to whiteness, and by withdrawing the points from each other a constant discharge took place through the heated air, in a space equal at least to four inches, producing a most brilliant ascending arch of light, broad, and conical in form in the middle. When any substance was introduced into this arch, it instantly became ignited; platina melted as readily in it as wax in the flame of a common candle; quartz, the sapphire, magnesia, lime, all entered into fusion; fragments of diamond, and points of charcoal and plumbago, rapidly disappeared, and seemed to evapo-

141. Dr. Werner Siemens was the first to employ (in 1847) gutta-percha as a means of insulating subterranean telegraph-wires. On the 18th of January, 1850, in a paper communicated to the Physical Society of Berlin, he stated that a subterranean wire covered with gutta-percha, and surrounded by the moisture of the earth, behaved like a colossal Leyden-jar. He also found that ordinary telegraph-wires charged themselves, though in a much smaller degree than the subterranean wires.

142. In 1838 Faraday predicted the retardation of the electric discharge by its own inductive action. ("Experi-

rate in it, even when the connection was made in a receiver exhausted by the air-pump; but there was no evidence of their having previously undergone fusion.

"When the communication between the points positively and negatively electrified was made in air, rarefied in the receiver of the air-pump, the distance at which the discharge took place increased as the exhaustion was made, and when the atmosphere in the vessel supported only one-fourth of an inch of mercury in the barometrical gauge, the sparks passed through a space of nearly half an inch; and by withdrawing the points from each other, the discharge was made through six or seven inches, producing a most beautiful coruscation of purple light, the charcoal became intensely ignited, and some platina wire attached to it, fused with brilliant scintillations, and fell in large globules upon the plate of the pump. All the phenomena of chemical decomposition were produced with intense rapidity by this combination. When the points of charcoal were brought near each other in non-conducting fluids, such as oils, ether, and oxymuriatic compounds, brilliant sparks occurred, and elastic matter was rapidly generated; and such was the intensity of the electricity, that sparks were produced, even in good imperfect conductors, such as the nitric and sulphuric acids.

"When the two conductors from the ends of the combination were connected with a Leyden-battery, one with the internal, the other with the external coating, the battery instantly became charged, and on removing the wires, and making the proper connections, either a shock or a spark could be perceived; and the least possible time of contact was sufficient to renew the charge to its full intensity."

mental Researches," 1333. "Faraday as a Discoverer," new edition, p. 89.)

143. In 1854 Faraday experimented with cables at the gutta-percha works of the Electric Telegraph Company. One hundred miles of gutta-percha covered wire were immersed in water, and a second hundred miles of a similar wire were placed in a dry tank. We will call the former the water wire, and the latter the air wire.

144. Connecting one pole of a battery with the earth, and connecting the other pole with one of the two insulated ends of the water wire, on breaking the connection and touching the wire a powerful shock was received; the discharge from the wire was also competent to ignite a Statham fuze. When, after having been in contact with the battery, the wire was separated and connected with a galvanometer, the instrument was powerfully affected.

145. A rush of electricity into the wire was declared by the galvanometer when contact was made; a rush out of the wire was declared when the wire between the battery and the galvanometer was connected with the earth. None of these effects were observed with the 100 miles of air wire.

146. Faraday, like Werner Siemens, rightly explained the effect by likening the cable to an enormous Leyden-jar, the wire constituting the interior, the water the exterior coating, with the gutta-percha insulator between them. In fact, the surface of the wire in these experiments amounted to 8,300 square feet, while the surface of the outer coating of water was 33,000 square feet. To the charge and discharge of this apparatus the effects observed were due.

147. In a subterranean line of telegraph 1,500 miles long were placed three galvanometers: one, a, at the beginning of the wire; a second, b, in the middle; and a

third, *c*, at the end, which was also connected with the earth.

148. Connecting the battery with the wire of the galvanometer *a*, that instrument was instantly affected; after a sensible time *b* was affected; and after a still longer time, *c*. It required, in fact, two seconds for the electric stream to reach the last instrument.

149. All the instruments being deflected, when the battery was suddenly cut off at *a*, that instrument instantly fell to zero, *b* fell subsequently, and *c* after a still longer interval.

150. By a brief touch of the battery-pole against *a*, that instrument was deflected, and could be allowed to fall back into its neutral condition before the electric power had reached *b*; *b* in its turn would be affected, and left neutral before the power had reached *c*.

151. In this case a wave of force was sent into the wire which gradually travelled along it, appearing in different parts of the wire at successive intervals of time.

152. It was even possible, by adjusted touches of the battery, to make several successive waves coexist in the wire.

153. When, after making and breaking contact at *a*, that galvanometer was connected with the earth, part of the electricity sent into the wire returned, and deflected *a* in the reverse direction; here currents flowed in opposite directions out of both extremities of the wire.

154. These effects of induction enabled Werner Siemens and Faraday to explain the widely-different velocities assigned by different experimenters to the electric current.

155. To pass through any conductor electricity requires time, the time being directly proportional *to the length* of the conductor.

156. But in the case of a submarine cable another cause of retardation comes into play, namely, the charging of the cable; the retardation here is proportional *to the square of the length* of the cable.

Artificial Cables.

157. It was to illustrate points like these and to determine the dimensions to be given to the Atlantic cables, that Mr. Cromwell Varley devised his artificial cables.

158. In one of these cables a resistance equal to that of a real cable 14,000 miles in length is obtained by introducing into the path of the current feebly-conducting liquids instead of metallic wires. The inductive action is obtained by means of condensers of tin-foil. In another artificial cable coils of wire are employed to give the necessary resistance.

159. The arrangement described in Note 81 is a condenser. But those constructed by Mr. Varley are of enormously greater area, the condensing sheets being separated from each other not by plates of glass, but by thin sheets of paper and paraffine. The vastness of the area and the proximity of the inducing surfaces combine to exalt the effect.

160. When the condensers themselves are charged by a battery, on discharging them they exhibit phenomena similar to those of a Leyden-jar. The shock, spark, and other effects of frictional electricity, are readily obtained.

161. A series of 50 condensers, for example, joined "in cascade," that is to say, with the outer coating of each joined to the inner coating of the next, when charged with a battery of 1,000 cells, yield powerful sparks, and deflagrate wires.

162. If the wire be bent and introduced into a glass of water, the glass is shattered by the discharge.

163. In the 14,000-mile artificial cable are introduced a series of eleven tubes containing the resisting liquid. Into these dip wires. One end of the charging battery is connected with the earth, and the other end can, at will, be connected with the artificial cable. A series of ten galvanometers are placed between the resisting tubes along the artificial cable.

164. When no condensers are employed, on making connection with the battery all the galvanometers appear to be simultaneously deflected.

165. When the condenser is introduced between each pair of resisting cells—ten condensers in all—the current has to charge each condenser to a certain degree before it can sensibly affect the galvanometer beyond the condenser. Hence, when the condensers are attached, the action on the galvanometers is successive, not contemporaneous.

166. Mr. Varley supposed his 14,000-mile artificial cable divided into sections representing stations in London, at Gibraltar, Malta, Suez, Aden, Bombay, Calcutta, Rangoon, Singapore, Java, and Australia. Supposing an actual cable laid, and galvanometers placed at these stations, the deflections obtained on establishing battery contact would be successive. They are represented by the deflections of the galvanometers associated with the artificial cable.

167. By varying the resistance and the amount of inductive condenser-surface, a representation of any other cable may readily be produced.

168. Connected with the needle of each of the ten galvanometers is a reflecting mirror, from which a brilliant spot of light is cast upon a screen. When the cable is not in action, the ten spots form a row along the same vertical line; when the battery contact is made, the successive

deflections of the galvanometers is declared by the successive motion of the spots.

Sketch of Ohm's Theory and Kohlrausch's Verification.

169. I have already spoken (Note 110) of the force which urges forward the electric current (the electromotive force). The amount of this force may be deduced from the action of the current, when opposed by different resistances, upon a freely-suspended magnetic needle.

170. If the wire which carries the current be cut across, the current ceases to flow. The electricity ceases to be *dynamic.* But at the two ends of the severed wire we have *static* electricity.

171. By suitable instruments the amount of this statical charge may be determined; it increases with the number of elements of the battery.

172. It is, moreover, proportional to the strength of the current obtained when the wires are reunited.

173. In this way the statical charge becomes a measure of dynamical action: electricity at rest is connected with electricity in motion.

174. In experiments on the electroscopic properties of the voltaic circuit it is necessary that the battery should be well insulated.

175. If the middle point of a wire which connects the two poles of a voltaic battery be connected with the earth, the tension of that point is null. The circuit gradually rises in tension right and left to the two poles of the battery. But on one side of the point we have exclusively positive electricity, while at the other side we have exclusively negative electricity.

176. At equal distances, at opposite sides of the zero-point, the tension is the same.

177. If any other point than the middle be connected

at the earth, it becomes the zero-point, right and left of which as before we have the two opposite electricities.

178. If the negative end of the battery be connected with the earth, the whole wire shows positive electricity; if the positive end be connected with the earth, the whole wire shows negative electricity.

179. The wire offers a certain resistance to the passage of the current. The battery itself is also in the circuit, and the current has to overcome its resistance also. But the resistance of the battery may be expressed by a certain length of the external wire. When this is done the sum of the lengths of both wires is called the *reduced length* of the circuit.

180. Given the reduced length of the circuit and the electro-motive force, we can determine by a simple calculation the electric tension of every point in the circuit.

181. The circuit through which the current flows may be represented by a horizontal line (called an abscissa); the electric tension at every point of the circuit may be represented by a vertical line (called an ordinate). If ordinates be drawn to represent the electric tensions at a great number of points of the circuit, the line joining the ends of all the perpendiculars will represent the distribution of electric tension in the circuit. The *steepness* of this line also represents what Ohm called the *electric fall*.

182. More strictly, the electric fall is the decrease in the length of the ordinate for the unit of length of the abscissa.

183. The total charge of the wire is expressed by the area of the triangle enclosed by the ordinate, abscissa, and line of fall.

184. The laws of the voltaic circuit as enunciated by Ohm, have been verified everywhere. The electroscopic

SKETCH OF OHM'S THEORY.

state of the circuit has been examined by Kohlrausch, and found to be in strict accordance with Ohm's theory.

185. Ohm assumed the passage of the electric fluid from one section to another of the connecting wire to be due solely to the difference of electric tension between the two sections; he further assumed the quantity of electricity transmitted to be proportional to this difference of tension, and from these fundamental assumptions he deduced the laws of the voltaic circuit.

186. These laws may be briefly stated thus:

a. The strength of the current is directly proportional to the electro-motive force.

b. The strength of the current is inversely proportional to the resistance.

c. If the wire which unites the two poles of battery be of the same material, and of the same thickness throughout, the "electric fall" is the same throughout the wire.

d. If the wire be of the same material, but of different thicknesses, the "fall" is steeper on the thin wire than on the thick. The "fall" is inversely proportional to the cross-section of the wire.

e. If the poles be connected by two wires of the same thickness, but of different resisting powers, the electric fall is steepest on the more resisting wire. The "fall" is directly proportional to the specific resistances of the wires.

187. In verifying these laws Kohlrausch employed a condenser to augment the feeble charges obtained from his voltaic cell, and he held this instrument to be essential. By an exceedingly skilful device Sir Wm. Thomson has rendered the condenser unnecessary, and has thus greatly simplified the means of demonstration.

Electro-chemistry.—*Chemical Actions in the Voltaic Cell: Origin of the Current.*

188. Philosophers suppose matter to be made of elementary parts called atoms, which are practically indivisible.

189. The elementary atoms can be caused to unite to form compound atoms, which are called molecules.

190. Thus water is formed of the combination of the atoms of oxygen and hydrogen; common salt is formed of union of atoms of chlorine and sodium; potash is formed by the union of the atoms of potassium and oxygen; the sulphuric acid also which we employed to acidulate our water is formed by the union of atoms of sulphur with atoms of oxygen.

191. When, as in our first experiment, two strips of zinc and platinum are dipped into acidulated water, the zinc, as we know, exerts a very strong attraction on the oxygen of the water. When the strips are united this attraction triumphs; the oxygen unites with the zinc, and a voltaic current is established.

192. The oxide of zinc here formed combines with the sulphuric acid and forms sulphuric zinc.

193. By this removal of the oxide from its surface the zinc is kept constantly clean, and thus enabled to attract other atoms of oxygen from the surrounding liquid. During this process the zinc gradually dissolves, and as long as this continues the electric current will flow. In fact, it is the constant dissolution of the zinc that maintains the permanent current.

194. The hydrogen of the water, as we have seen, escapes as a free gas from the surface of the platinum, which, unlike the zinc, is not dissolved.

195. We are not yet quite clear as to the precise way

in which the electric current is supported by the solution of the zinc, but the following facts and speculations ought to be known to you.

196. When two different metals are brought into contact, with no liquid between them, one of them charges itself with positive and the other with negative electricity. We have here the famous "contact force" which Volta and his followers considered to be the urging power of the voltaic current.

197. But the generation of heat, and the performance of mechanical work, by the mere contact of two metals, would be equivalent to a perpetual motion. It would be at variance with the law which requires for the production of any power an equivalent consumption of some other power.

198. It is, however, a fact that when two different metals touch each other the positive electricity resorts by preference to one metal, and the negative electricity to the other; the two electricities are as it were attracted differently by the two metals.

199. This difference of attraction, however, only causes a momentary rearrangement of the two electricities, which pass, when the contact is made, into a new condition of equilibrium. As long as the contact continues this equilibrium is not disturbed; there is no continuous current.

200. We may regard the distinct atoms which enter into the molecules of a compound as charged in a similar manner. For example, the atoms of hydrogen and oxygen when they unite to form a molecule of water, may be looked upon as charged like the two touching metals. This would be the case if the atoms, like the metals, possessed different attractions for the two electricities.

201. When strips of zinc and platinum are plunged

in such a liquid, the positively-charged atom will turn toward the one metal, and the negatively-charged atom toward the other.

202. But, unless the metals touch each other, electrical equilibrium immediately sets in, a constant state of electric tension being set up at the free ends of the two metals.

203. The electricity at the ends may be permitted to flow into a condenser, and may be thus stored up; such a condenser may thus be discharged through a covered wire which passes round a magnetic needle, a deflection of the needle being thus produced.

204. Thus in Davy's experiment with his large voltaic battery, wherewith he charged his battery of Leyden-jars, the latter, after having been charged, might be discharged through a galvanometer, a magnetic deflection being thus produced.

205. But the metals, once relieved of their charge, would immediately reload themselves with electricity, and might be again employed to charge a Leyden battery, and to produce a deflection of a magnetic needle.

206. At no moment during this process the battery circuit would be complete; still we should have a succession of magnetic actions similar to those observed with a closed circuit.

207. In fact, in the closed circuit the solution of the zinc incessantly removes the charged surface of that metal by dissolving it away, and enables the zinc to take a fresh charge; an incessant effort, never fully satisfied, is made to establish electric equilibrium; the incessant renewal of the effort maintains the electric current.

Chemical Actions at a Distance: Electrolysis.

208. Thus, then, in the cell where the voltaic current is generated chemical action occurs. We have, on the one

hand, the decomposition of the water, and on the other the combination of the zinc with the oxygen and the sulphuric acid.

209. But a voltaic current can also produce chemical action at a distance from its place of generation. This discovery, as stated in Note 127, was made in the year 1800 by Nicholson and Carlisle.

210. We cannot decompose water by a single voltaic cell; but when two or more cells are united to form a battery, the current from such a battery, when sent through acidulated water, tears asunder the united atoms of oxygen and hydrogen.

211. The oxygen is set free at the place where the current enters; the hydrogen is set free at the place where the current quits the liquid. If the direction of the current be reversed, the oxygen and hydrogen instantly change places.

212. It must be clearly borne in mind that the direction of the current, as already defined, is the direction in which the positive electricity moves. Knowing, therefore, the places at which the oxygen and hydrogen are liberated, we can infer with certainty the direction of the current through the liquid.

213. For every volume of oxygen liberated in the decomposition of water by a voltaic current, two volumes of hydrogen are set free.

214. Electro-chemical decomposition is called *electrolysis;* and the compound liquid decomposed by the electric current is called an *electrolyte.*

215. The electric current formed a powerful means of analysis in the famous experiments of Sir Humphry Davy in 1807.

216. By operating with the current upon ordinary potash, Davy found the base of this substance to be a

metal of exceeding lightness, and with an extraordinary appetite for oxygen. When placed on water, it floated on the liquid, and combined with its oxygen. By the heat thus generated the liberated hydrogen was caused to burst into flame. When a globule of the metal was placed on ice, it burned with a bright flame, and the hole made by the heat was filled with a solution of potash.

217. Soda, treated in the same manner, also yielded a metal resembling that of potash. Thus Davy, by the use of the voltaic current, decomposed the alkaline earths, and greatly expanded our knowledge of chemistry.

218. To obtain these effects it is necessary to bring the potash and the soda to a state of fusion by heat. In the solid state they are non-conductors of electricity. In fact, the molecules, when rigid, cannot turn in the manner indicated in Note 201. To conduct the current, it is necessary that they should thus turn and be decomposed.

219. When the current is sent through a solution of common salt, it decomposes both the water and the salt. The chlorine of the salt, in company with the oxygen of the water, appears where the current enters the liquid. The sodium of the salt, in company with the hydrogen of the water, appears where the current quits the liquid.

220. Chlorine possesses powerful bleaching properties; and if the solution of salt be colored with indigo or litmus, the presence of the chlorine is declared by the destruction of the color.

221. When a current is sent through a solution of iodide of potassium, the brown substance iodine is set free where the current enters, while the metal potassium is set free where the current quits the solution. The experiment may be made by moistening bibulous paper with the dissolved iodide.

222. In electrolysis it is usual to immerse two plates of

platinum, or of some other suitable substance, in the liquid to be decomposed, and to send the current from plate to plate. The plate at which the current enters the liquid is called the Positive Electrode, the plate at which the current quits the liquid is called the Negative Electrode. Without the liquid these electrodes would, as we have already learned, charge themselves with positive and negative electricity.

223. But inasmuch as electricities which attract each other are of opposite qualities, the substance which is liberated at the positive electrode is called the Electro-Negative constituent, while the substance liberated at the negative electrode is called the Electro-Positive constituent of the liquid.

224. Thus, in the examples above given the oxygen, chlorine, and iodine, are the electro-negative elements; the hydrogen, sodium, and potassium, being the electro-positive elements.

225. The terms electro-positive and electro-negative are, however, relative, for a substance may be electro-positive in one combination, and electro-negative in another.

226. If an electric current be conducted through a solution of sulphate of soda, it separates the sulphuric acid from the soda; the presence of the acid may be proved by its turning a vegetable color red.

227. When nitrate of silver or acetate of lead is decomposed by a voltaic current, crystals of silver, or of lead, are deposited on the negative electrode.

228. The chemical actions of the electric current, some examples of which are here given, constitute what is called Electro-chemistry.

229. Electro-plating and gilding and the electrotype process are important applications of electro-chemistry.

Here a chemical compound containing gold, silver, or copper, is decomposed by a voltaic current, the metal being deposited on the surface intended to be coated with it.

230. If the surface on which the metal is deposited have a design engraved upon it, the lines of the engraving are accurately filled by the metal which, when the deposit is thick enough, may be detached, a perfect copy of the design being thus obtained.

Measures of the Electric Current.

231. The *tangent-compass*, devised by Weber, consists of a vertical ring of brass or copper, in the centre of which swings a small compass-needle. The ring being placed in the magnetic meridian, the needle is deflected when a current is sent round the ring. The strength of the current can be proved to be proportional to the tangent of the angle of deflection; hence the name of the instrument.

232. The *voltameter* is an instrument devised by Faraday to measure the strength of an electric current. It consists of a graduated tube which receives and measures the quantity of gas generated by the current in a given time.

233. The strengths of a series of currents measured by the voltameter are accurately proportional to the same strengths measured by the tangent-compass. Placing a tangent-compass and a voltameter in the same series of circuits, the tangents of the angles observed in the one case are accurately proportional to the quantities of gas generated in the other.

Electric Polarization: Ritter's Secondary Pile.

234. When an electric current is sent through acidulated water a film of oxygen covers the positive electrode, and a film of hydrogen covers the negative electrode. One of these two substances being electro-positive, and the other electro-negative, they act in the liquid like two different metals; the hydrogen plays the part of zinc, and the oxygen plays the part of platinum.

235. Interrupting the primary battery circuit, and uniting together the two plates covered with their respective films, an electric current is obtained.

236. The direction of this current is from the hydrogen film to the oxygen film in the liquid, and from the oxygen film to the hydrogen film through the connecting wire.

237. Two electrodes thus covered with condensed gaseous films are said to be *polarized;* and the currents obtained from them are called currents of polarization.

238. Now the battery current being always from oxygen to hydrogen (see Note 211), it is plain that the current of polarization is always opposite in direction to the battery current employed to polarize the electrodes.

239. When a decomposition cell with platinum plates is introduced into a voltaic circuit, it is found that the battery current, though strong at starting, gradually sinks. This sinking is due to the gradual development of the antagonistic current of polarization.

240. Also in the cells of the battery itself this current of polarization may come prejudicially into play. When two metals, say zinc and platinum, and one liquid, say acidulated water, are employed, the platinum plate is coated with a film of hydrogen.

241. This hydrogen, being electro-positive, resembles

a plate of zinc, so that when it is present we have, as it were, zinc opposed to zinc in the battery.

242. Were both plates actually of zinc, we could have no current; and with the hydrogen film which approximates to zinc we have only a feeble current. To get the full effect of the zinc and platinum some means must be devised to remove from the platinum its film of hydrogen.

243. This is effected in Grove's battery by the employment of *two liquids*. The one is strong nitric acid, which contains the plate of platinum; the other is dilute sulphuric acid, which contains the plate of zinc. The nitric acid is placed in a vessel of porous earthenware, which becomes saturated with the liquid and allows the current to pass through it.

244. When the current passes, the hydrogen liberated at the platinum electrode in Grove's cell is instantly oxidized by the nitric acid, and prevented from forming a film upon the surface of the platinum.

245. If instead of employing a single decomposition cell and a single pair of platinum electrodes, we employ a series of such cells, and send the same current through them all, we convert every pair of such plates into an active voltaic couple; and if the number of such couples be great, effects of great intensity may be obtained.

246. If instead of using decomposition cells we simply employ a series of plates of the same metal, say a series of half-crowns, separated from each other by pieces of bibulous paper or by bits of cloth wetted with acidulated water; on sending a voltaic current through such a pile of plates, we liberate on one of the surfaces of each plate a film of oxygen, and on the other surface a film of hydrogen. These play the part of the two different metals in the pile of Volta.

247. The electro-motive force of such a pile may be far greater than that of the battery which charges it. It may produce a far more brilliant spark, and urge its current against resistances which would be quite insuperable to the original battery current.

248. The discoverer of this form of pile was Ritter; it is sometimes called *the secondary pile*, to distinguish it from the battery which charges it.

Faraday's Electrolytic Law.

249. When the self-same current is sent through a series of cells containing various compound liquids, the same amount of liquid is not decomposed in all cases.

250. Let the current be sent in succession through a series of cells containing water, oxide of lead, chloride of lead, iodide of lead, and chloride of silver; then taking them in the above order, the weights of the liquids decomposed are represented by the numbers 9, 111.5, 139, 230.5, 143.5.

251. The question now is, how are these weights of the respective substances divided between the two electrodes? Supposing the numbers to express grains, we should have the following division between the electrodes:

At the positive electrode.			At the negative electrode.	
Water.........	8	grains oxygen....	1.	grain hydrogen.
Oxide of lead...	8	" "	103.5	grains lead.
Chloride of lead.	35.5	" chlorine...	103.5	" "
Iodide of lead..	127	" iodine	103.5	" "
Chloride of silver	35.5	" chlorine...	108	" silver.

252. Now these numbers express the combining proportions of the respective substances; by the electric current in all cases the law of combination as regards quantity is exactly inverted. The substances combine in equivalent proportions; they are decomposed in precisely

the same proportions. This is the celebrated law of electrolysis discovered by Faraday.

253. In no case in the body of the electrolyte is any decomposition observed; in no case is any gas there liberated. The substances set free appear at the electrodes, and there alone.

254. Taking water as an illustration, the process is to be figured thus: When the electrodes, charged with electricity from the battery, are plunged into the liquid, the oxygen atom of the water turns toward the positive, and the hydrogen atom toward the negative electrode.

255. If the electro-motive force be strong enough, the oxygen is torn away from its hydrogen; the free hydrogen immediately converges its attraction on the next adjacent oxygen atom, and unites with it, dislodging at the same time the hydrogen with which that atom had been previously combined. Another atom of hydrogen is thus liberated, which in its turn decomposes the adjacent water-molecule. Thus through the chain of molecules run a series of decompositions, followed by immediate recompositions, until the negative electrode is reached. Here the hydrogen, having no further oxygen with which to combine, is liberated as a gas. This is the theory of Grotthuss, which at all events fairly embraces the facts.

Nobili's Iris Rings.

256. The hardness of steel in tempering it is judged by its color, which is due to a film of oxide overspreading the steel. The oxide which forms on the surface of molten lead also shows vivid colors.

257. These are the colors of *thin plates* investigated by Newton and explained by Thomas Young.

258. By electro-chemical decomposition Nobili pro-

duced such colors in a very beautiful manner. Placing, for example, a polished steel plate in a dilute solution of acetate of lead, and connecting the plate with the positive pole of a voltaic battery, on dipping the end of a wire connected with the negative pole into the solution, the peroxide of lead is liberated on the surface of the steel immediately under the wire; and a film gradually diminishing in thickness spreads from that point outward. Round this point we have a series of concentric circles showing vivid iris colors.

259. These colors, like all those of thin plates, depend upon the thickness of the film, which diminishes as the distance traversed by the current increases.

(Du Bois-Reymond has shown that when the point from the negative end of the battery is very near the steel plate, the thickness of the film corresponding to the different circles is inversely proportional to the cubes of their radii.)

Distribution of Heat in the Circuit.

260. When the two ends of a voltaic battery are connected by a thick wire of good conducting material the wire is not sensibly heated; the heat due to the oxidation of the zinc is in this case confined to the battery itself.

261. But if the two ends of the battery be connected by a wire that offers a resistance to the current, the wire is heated, and may, if properly chosen, be raised to a white heat.

262. Considering the battery as the hearth where the zinc is burnt, we might be led to infer that the heat due to the combustion of the zinc is liberated on the hearth itself, and that its amount depends solely upon the quantity of zinc consumed.

263. This, however, is not the case. Let the battery,

with its two ends united by a thick wire, be surrounded by a vessel of water, to which the heat developed by the oxidation say of an ounce of zinc is communicated; the quantity of heat developed is measured by the rise of temperature of the water.

264. Let the battery, with its two ends united by the resisting wire, be placed in the same vessel, and let the heat generated in the battery by the oxidation of an ounce of zinc be again determined; this heat will be less than that observed in the last experiment.

265. If the connecting wire be now enclosed in a separate vessel, and if the heat generated in the wire be thus determined, on adding this amount of heat to that liberated in the battery, a total heat is obtained exactly equal to that generated in the battery alone, when the good conducting wire was employed.

266. In fact, the absolute amount of heat generated by the oxidation of an ounce of zinc is perfectly constant; but it may be distributed in various proportions between the battery and the external circuit.

Relation of Heat to Current and to Resistance.

267. On what does heat developed in a wire uniting the two ends of a voltaic battery depend?

268. It depends, in the first place, on the strength of the current, but it is not simply proportional to that strength.

269. Let the strengths of a series of currents, determined either by the tangent-compass or the voltameter, be represented by the numbers 1, 2, 3, 4, then the quantities of heat developed in the same wire by these respective currents are expressed by the numbers 1, 4, 9, and 16.

270. The heat generated is therefore proportional to *the square* of the strength of the current.

271. Preserving the strength of the current constant, the heat generated is proportional to the electrical resistance of the wire through which it passes. These important principles were established by Joule.

272. Thus if one of two equal currents pass through a silver wire, and the other through a platinum wire of the same length and thickness, the heat generated in the platinum will be ten times that generated in the silver, because the resistance of the former is ten times that of the latter. To urge the current through the platinum in this case would, however, require greater battery-power than that necessary for the silver.

273. Hence, when *the same current* is sent through a wire composed of alternate lengths of silver and platinum of equal thickness, the platinum spaces may be raised to a white heat, while the silver is not raised to the faintest glow.

Magneto-Electricity: Induced Currents.

274. In a conductor near to, but not in contact with a voltaic circuit, a current is aroused when the circuit is *established*. When the circuit is *interrupted* a current is also aroused in the conductor.

275. Thus, supposing the voltaic circuit to be bent into the shape of a ring; and that a second ring, not in the circuit, is placed near the first: at the completion, and at the interruption of the circuit, a current will run round the second ring.

276. The two currents in the second ring are called *secondary* currents. They are of momentary duration. They impart, in passing, a shock to a magnetic needle round which they are sent, and by the motion of which

their existence is demonstrated. But they vanish immediately, being quenched by the resistance of the ring and converted into heat.

277. These two momentary currents flow in opposite directions through the ring. The secondary current, excited on making the circuit, is opposed in direction to the primary exciting current; that started on interrupting the circuit flows in the same direction as the primary.

278. These secondary currents are called *induced currents*. They were discovered by Faraday in 1830, and described by him in his Philosophical papers for 1831.

279. If, instead of employing a single ring, we make use of an electro-magnetic helix, every coil of the helix will furnish its quota of current, and the sum total of effect is much greater than when only a single ring or coil is employed.

For the following experiments, two flat spirals, each formed of covered copper wire, are used.

280. One of the spirals is laid flat upon a table, its two ends being connected with a galvanometer; the other spiral is connected with a voltaic battery, with which the connection can be established or broken at pleasure. Let us call this the *inducing* or *primary* spiral, and that connected with the galvanometer the *secondary* or *induced* spiral.

281. Laying one spiral upon the other, on sending a current through the primary, the needle of the galvanometer is suddenly driven aside by the current induced in the secondary; but the force which acts upon the needle passes away in an instant, the needle returning to its first position.

282. On interrupting the current the needle also receives a shock, being deflected in the opposite direction. It thus declares the existence of a second temporary cur-

MAGNETO-ELECTRICITY. 183

rent in the secondary spiral. The directions of these two currents, with reference to that of the primary, have been already indicated; Note 277.

283. Holding the secondary spiral at a distance from the primary with the current flowing through the latter; on causing the secondary spiral *to approach* the primary, a current is aroused; this current ceases the moment the motion toward the primary ceases.

284. On *withdrawing* the secondary spiral from the primary, a current is also aroused; this current also ceases the moment the motion of withdrawal ends.

285. The current excited by approach is opposed in direction to the primary; the current excited by withdrawal is in the same direction as the primary.

286. Two electric currents flowing in the same direction attract each other; if they flow in opposite directions they repel each other.

287. Hence, to make the secondary spiral approach its primary, we have to overcome a *repulsion;* while to withdraw the secondary from the primary we have to overcome an *attraction.* Thus in order to produce these induced currents *we must expend mechanical force.*

288. The force thus expended appears as heat in the secondary wire after the cessation of the induced current. It is the mechanical equivalent of that heat.

289. The approach of a magnetic pole to the secondary spiral and the withdrawal of the pole from the same spiral also arouse induced currents. But, as before, it is only during the periods of approach and withdrawal that the current appears.

290. Thus by the mere motion of a magnet, and without any battery or machine, electric currents may be produced.

291. Every change of the magnetic condition of the

space near a secondary coil, or within it, produces an induced current in the coil. If the change be an augmentation of magnetism, the current is in one direction; if it be a diminution of magnetism, the current is in the opposite direction.

292. When a long secondary coil surrounds a primary coil with a core of iron, by breaking and making the circuit of the primary in rapid succession, a series of powerful discharges may be obtained. An automatic apparatus is usually employed to make and break the circuit.

293. Such Induction Coils have been constructed with great skill by Ruhmkorff, and are, therefore, sometimes called Ruhmkorff's coils. Mr. Apps has recently produced induction coils of astonishing power.

294. The power of a coil depends mainly on the perfection of the insulation of its coils. The induced currents in a Ruhmkorff's coil may possess thousands of times the electro-motive force of the primary which excites them. They are able, for example, to overleap as sparks, distances thousands of times greater than that possible to the primary.

Relation of Induced Currents to the Lines of Magnetic Force. Rotatory Magnetism.

295. The foregoing phenomena and principles were all laid bare by Faraday. He also established most important relations between his induced currents and the lines of force surrounding a magnet. See Note 25.

296. He proved that when a conductor moves *along* the lines of force no induced currents appear; but that when it moves *across* the lines of force such currents are generated.

297 He proved, for example, that when a metal disk

is caused to rotate so as to be tangent to the lines of force, no current appears; while when the disk, in its rotation, *cuts* the lines of force, currents flow along the disk, from the centre to the circumference and from the circumference to the centre. Closed circuits are thus established in the disk.

298. This, in fact, is the "Magnetism of Rotation," discovered by Arago in 1820, which received complete explanation at the hands of Faraday.

299. Faraday showed that the lines of force of terrestrial magnetism suffice to produce induced currents when they are intersected by the rotating disk. In fact, all the effects of magneto-electric induction may be obtained from the magnetism of the earth.

300. When a conductor rotates round an axis which is parallel to the lines of force, it experiences simply the resistance due to the friction of the air; but if the axis of rotation be transverse to the lines of force, the rotation is retarded by the interaction of the magnet and the induced currents.

301. This retardation may become so powerful as instantly to arrest the rotation. If, for example, a cube or sphere of copper suspended from a twisted string be caused to spin, by untwisting, between the poles of an unexcited electro-magnet, it experiences the retardation due to air friction only; but on the supervention of the magnetic force the rotation is suddenly arrested. Faraday also showed that in passing a plate of copper rapidly to and fro between the magnetic poles you seem to be cutting cheese, though nothing is visible. It is as if pure space were a kind of solid.

302. If by mechanical means the conductor be compelled to rotate or to move to and fro between the excited poles, it will be heated. Joule first demonstrated this

but a very striking demonstration of it was given by Foucault, who heated his celebrated gyroscope in this way. The heat is readily rendered sufficiently intense to melt fusible metal. Between the unexcited poles no effect of this kind is produced.

303. The repulsion set up by induced currents between the helices and the moving masses of iron in an electro-magnetic engine, would of itself limit the practical application of electricity as a motive power. Nevertheless, though such engines speedily reach the limit of their action, the conversion of molecular force into mechanical effect may be rendered far more perfect than in the case of the steam-engine.

The Extra-Current.

304. If the secondary coil of a Ruhmkorff's machine have its ends united, the secondary circuit being then complete, the spark obtained in breaking the primary is small. On separating the two ends of the secondary the primary spark is instantly augmented.

305. The diminution of the spark is due to the reaction of the completed secondary circuit upon the primary. When the secondary circuit is interrupted this reaction ceases.

306. The primary circuit in its turn can, when complete, react upon the secondary. It *is* complete whenever contact is *made* by the automatic contact-breaker. A great enfeeblement of the secondary current is the consequence. When the primary circuit is interrupted the reaction does not exist; there is no enfeeblement, the full power of the secondary being developed. It is on this account that in Ruhmkorff's coil we obtain discharges *in a single direction* only, instead of discharges alternating in direction.

THE EXTRA-CURRENT.

307. The reaction here referred to connects itself with what is called the *extra-current*.

308. When a current is sent through a single primary coil, the primary current excites in the wire which carries it, a secondary current opposed in direction to the primary. The primary arouses an antagonist in its own path, which, however, immediately disappears.

309. When the primary circuit is broken, a secondary current of momentary duration, and having the same direction as the vanishing primary, is evoked in the coil.

310. Each of the two currents evoked *in the primary circuit itself*, at the commencement and at the cessation of the primary current, has been called by Faraday an *extra-current*.

311. The spark obtained on breaking the primary circuit is augmented in brilliancy and power by the extra-current.

312. If a second circuit be associated with the primary; if, for example, two covered wires are wound round the same reel; on making one of them a primary circuit we have the brilliant spark due to the extra-current, *as long as the ends of the second coil remain unconnected.*

313. But the moment they are connected the extra-current in the primary circuit disappears; there is an instant reduction in the brilliancy of the spark.

314. This is an example of the reaction referred to in Note 304. By the closing of the secondary circuit the extra-current is formed *in it* instead of in the primary one. Here, in fact, the extra-current becomes an ordinary induced current; it is only so long as it remains in the primary circuit that its distinctive name is applied to it.

Influence of Time on Intensity of Discharge. The Condenser.

315. The intensity of the secondary current—its "discharging distance," for example—depends upon the rapidity with which the primary is interrupted.

316. I have already referred to the passage of particles between the two severed terminals of a circuit. By these particles the current may be kept up for a short time after the terminals have been disunited. A *gradual* dying away of the primary is the consequence.

317. But to produce the maximum secondary intensity it is necessary that the primary should be extinguished *at once*.

318. This is very effectually accomplished if the primary be broken between the poles of a strong magnet. The secondary spark may be thus made to overleap distances, vast in comparison with those possible to it when the rupture of contact occurs far away from the magnetic poles.

319. The magnet quenches immediately the stream of particles which accompany the spark. Thus, instead of being spread over a very sensible interval, the whole power of the primary is concentrated into an instant of time.

320. This concentration is announced by the loudness of the report of the primary spark. This augmentation of loudness was first observed by Page; it was explained by Rijke, who also exalted in the way here indicated the discharge of the secondary coil.

321. The injurious effect of the spark produced by the rupture of contact in Ruhmkorff's coil is much diminished by the employment of a condenser, which is attached to the primary. It was introduced by Fizeau.

Electric Discharge through Rarefied Gases and Vapors.

322. The electricity from the prime conductor of an electrical machine passes through the air in the form of a dense and brilliant spark, which produces a very audible report.

323. When the discharge passes through rarefied air the discharging distance is augmented, and by sufficiently rarefying the air the discharge may be caused to pass *silently*. It then fills the tube through which it passes with a rosy light.

324. This rosy light has the same origin as that of the Aurora Borealis; it is due to the nitrogen of the air.

325. Every attenuated gas has its own characteristic color when traversed by the electric discharge. When examined by a prism the color resolves itself into distinct bands; the nature of the gas may, indeed, be inferred from the analysis of its spectrum.

326. The discharge of the induction coil through attenuated media produces luminous effects similar to those produced by the electric machine.

327. The tubes containing the attenuated gases or vapors are usually called *vacuum tubes*. Through the tubes pass platinum wires which are fused into the glass, and between which the discharge passes.

328. Such tubes are produced in great perfection by Geissler, of Bonn, and are sometimes called Geissler's tubes.

329. Under certain circumstances, the luminous discharge is composed of distinct luminous strata separated from each other by dark intervals transverse to the direction of the discharge. These strata were first observed by Grove; they were observed independently and finely developed by Ruhmkorff.

330. The luminous strata were believed to arise from the intermittent action of the contact-breaker of the induction coil; but Gassiot produced them both with the electric machine, and with his battery of 3,500 cells, where no contact-breaker is employed.

331. Every single discharge of the induction coil through a properly-chosen medium resolves itself into a series of pulses, which declare themselves as a stratified discharge. Under similar circumstances the discharge from the voltaic battery also is resolved into a series of pulses which are declared by their stratifications.

Action of Magnets on the Luminous Discharge.

332. The luminous discharge is to all intents and purposes an electric current, and is acted on by a magnet like a wire carrying a current.

333. But the flexibility of the luminous current in rarefied gases enables the magnet to act upon it in a manner peculiarly interesting and instructive.

334. Placing, for example, a tube through which the luminous discharge is passing between the poles of an electro-magnet, by exciting the magnet the stream of light may be either deflected or wholly extinguished.

335. In the latter case, by interrupting the current passing round the magnet, or by lifting the tube out of the magnetic field, the luminous discharge is restored.

336. In certain cases, when the luminous discharge consists simply of a feeble glow, the supervention of the magnetic force draws a series of strongly-illuminated strata from the positive terminal of the vacuum-tube; when the magnetism is interrupted, these strata retreat again in succession, as if swallowed up by the positive pole. A number of exceedingly beautiful experiments of this character has been made by Gassiot.

337. It has been stated in Note 306 that the discharges from the induction coil proceed always in the same direction; hence, in each vacuum-tube we have a positive terminal or pole, and a negative terminal or pole.

338. When the light surrounding the negative terminal is subjected to a magnet, it ranges itself exactly along the lines of magnetic force; the light at the positive terminal shows no such action. This discovery is due to Plücker.

Magneto-electric Machines. Saxton's Machine. Siemens's Armature.

339. Faraday's discovery of Magneto-electricity was announced in 1831. In 1833 a machine was constructed by Saxton for the more copious development of magneto-electric currents.

340. In it copper-wire coils, within which were placed cores of iron, were caused to rotate before the poles of a powerful magnet.

341. On the approach of a coil to one of the poles of the magnet, a powerful current, whose direction depended on the nature of the pole, was induced in the coil. When the coil retreated from the magnetic pole, a current in the opposite direction was induced. This production of opposite currents by approach and withdrawal has been already referred to in Notes 283, 284.

342. By means of an instrument called *a commutator*, which reversed one of the induced currents at the proper moment, the opposite currents were caused to flow in the same direction.

343. The cores of soft iron and their associated coils constitute what is called an *armature*. In Saxton's armature the coils were wound *transversely* to the iron cores.

344. But by winding his coils *longitudinally*, or parallel to the axis of the core, and placing the armature so formed between the poles of a series of horseshoe magnets, Siemens obtained magneto-electric currents much more powerful than those of Saxton.

Wilde's Machine.

Things were in this state when, in 1866, Wilde made an important addition to our knowledge of magneto-electricity.

345. He conducted the current obtained by means of Siemens's armature round an electro-magnet, and found that the magnetism thus excited was far greater than that of the entire series of steel magnets employed to generate the magneto-electric current.

346. Thus, in one case, he found that whereas the series of permanent magnets taken collectively was competent to support a weight of 40 lbs. only, the electro-magnet which they excited sustained a weight of 1,088 lbs.

347. To produce this effect, however, it was necessary that the armature of the magneto-electric machine should rotate with great rapidity.

348. But Wilde went farther. Forming his electro-magnet from a large plate of iron, and placing between its long poles a correspondingly long armature, similar in shape and construction to that of the magneto-electric machine, he obtained from this second armature currents of enormously greater power than those obtainable from the first.

349. These currents could in their turn be sent round a second electro-magnet, formed from a larger plate of iron. Furnished with a rotating armature, this second electro-magnet produced effects previously unknown. Rods of iron a quarter of an inch in thickness were fused by the

currents, and they were also found competent, when discharged between carbon terminals, to produce a light of intolerable brilliancy.

Siemens's and Wheatstone's Machine.

350. The next great step in magneto-electricity was made simultaneously by Dr. Werner Siemens and Sir Charles Wheatstone.

351. Expressed generally, this discovery consists in exalting, by means of its own action, to a high pitch of intensity an infinitesimal amount of magnetism.

352. Conceive an electro-magnetic core with a very small amount of residual magnetism, which is never wholly absent when iron has been once magnetized. Let a secondary coil, with cores of soft iron, rotate before the poles of such a magnet. Exceedingly feeble induced currents will circulate in the secondary coil. Let these induced currents, instead of being carried away, be sent round the electro-magnet which produced them; its magnetism will be thereby exalted. It is then in a condition to produce still stronger currents. These also being sent round the magnet, raise its magnetism still higher; a more copious production of induced currents being the consequence. Thus by a series of interactions between the electro-magnet and the secondary helix, each in turn exalting the other, the electro-magnet is raised from a state of almost perfect neutrality to one of intense magnetization.

353. When the magnet has been raised to this condition, other coils than those employed to magnetize it may be caused to rotate before, or between, its poles; the currents from these coils may be carried away and made use of, for magnetization, for chemical decomposition, or for the electric light.

354. The first magneto-electric machine used to pro-

duce a light sufficiently intense for light-houses was constructed by Mr. Holmes. In it permanent steel magnets and rotating helices were employed. Mr. Holmes has lately constructed a very powerful machine on the principle of Siemens and Wheatstone.

Induced Currents of the Leyden-Battery.

355. If a Leyden jar, or battery, be discharged through a primary spiral, it evokes a current in a secondary spiral. With a strong charge this secondary current may be caused to deflagrate a foot of thin platinum wire.

356. If the current from the secondary spiral be led round a third spiral which faces a fourth; on discharging the battery through the primary spiral, the secondary in the third spiral acts the part of a primary, and evokes in the fourth spiral a *tertiary current.*

357. With another pair of spirals this tertiary current can he made to generate a current of the *fourth order;* this again, with another pair of spirals, a current of the *fifth order.* All these currents can impart shocks, ignite gunpowder, or deflagrate wires.

For the investigation of the Induced Currents of the Leyden-Battery we are indebted to Prof. Joseph Henry, Director of the Smithsonian Institution, and to Prof. Riess, of Berlin.

THE END.

www.ingramcontent.com/pod-product-compliance
Lightning Source LLC
Chambersburg PA
CBHW030819190426
43197CB00036B/610